Climate Change, Agriculture and Rural Livelihoods in Developing Countries

Advances in Asian Human-Environmental Research

Aims and Scope

The series aims at fostering the discussion on the complex relationships between physical landscapes, natural resources, and their modification by human land use in various environments of Asia. It is widely acknowledged that human-environment-interactions become increasingly important in area studies and development research, taking into account regional differences as well as bio-physical, socio-economic and cultural particularities.

The book series seeks to explore theoretic and conceptual reflection on dynamic human-environment systems applying advanced methodology and innovative research perspectives. The main themes of the series cover urban and rural landscapes in Asia. Examples include topics such as land and forest degradation, glaciers in Asia, mountain environments, dams in Asia, medical geography, vulnerability and mitigation strategies, natural hazards and risk management concepts, environmental change, impacts studies and consequences for local communities. The relevant themes of the series are mainly focused on geographical research perspectives of area studies, however there is scope for interdisciplinary contributions.

For further volumes:
http://www.springer.com/series/8560

Keshav Lall Maharjan • Niraj Prakash Joshi

Climate Change, Agriculture and Rural Livelihoods in Developing Countries

Springer

Keshav Lall Maharjan
Graduate School for International Development and Cooperation (IDEC)
Hiroshima University
Hiroshima, Japan

Niraj Prakash Joshi
Graduate School for International Development and Cooperation (IDEC)
Hiroshima University
Hiroshima, Japan

ISSN 1879-7180 ISSN 1879-7199 (electronic)
ISBN 978-4-431-54342-8 ISBN 978-4-431-54343-5 (eBook)
DOI 10.1007/978-4-431-54343-5
Springer Tokyo Heidelberg New York Dordrecht London

Library of Congress Control Number: 2013937295

Printed on acid-free paper

Springer is part of Springer Science+Business Media (www.springer.com)

Preface

Climate change has now become an indisputable fact. It will affect the economies of developed as well as developing countries. However, owing to the high dependency of developing countries, more specifically the Least Developed Countries (LDCs), on naturally sensitive sectors such as agriculture and forestry, these developing countries are already exposed to a higher risk from any unusual changes in climate phenomena compared to the developed countries. Thus, any adverse impact on these sectors will have a significant bearing on the livelihoods of the majority in such countries. Moreover, their geographical disadvantages as well as the lack of economic strength to cope with and adapt to such adverse changes put them into a highly vulnerable position. Therefore, this book is prepared as course material for the Global Environment Leaders Education Program for Designing a Low-Carbon Society (GELs Program), Graduate School for International Development and Cooperation (IDEC), Hiroshima University, in order to provide basic information on climate change and its relation to agriculture and rural livelihoods as well as information on international climate change regimes related to agriculture.

In considering the above-mentioned facts, this book discusses the impact of climate change in developing countries, taking the case of Nepal as an example. In doing so, in Chap. 1 the book starts with a basic understanding of climate change and the relation of agriculture to climate. Chapter 2 deals with the emission of greenhouse gases from various sectors of economic activities in different countries. The effect of several aspects of climate change on plant and animal physiology is discussed in Chap. 3. Agriculture is an important sector in developing countries, and the inclusion of this sector in international climate change negotiations will have an impact on the economy of these countries. Hence, Chap. 4 describes agriculture in international climate change negotiations. Similarly, Chap. 5 discusses cost and opportunities resulting from mitigation and adaptation in agriculture through international climate change negotiations. Chapter 6 highlights some important methodologies to assess the impact of climate change in agriculture. With this basic understanding of climate change, agriculture, international climate change regimes, and methodologies to assess impact of climate change in agriculture, Chap. 7 discusses the effect of climate change in regional agriculture production, food prices, and food insecurity.

The case of Nepal is highlighted to discuss the above issues, i.e., climate change and its relation to agriculture and rural livelihoods in a local context, to enhance the understanding of the location specificity of climate change and its impact on agriculture and rural livelihoods through adaptation, mitigation, and resilience leading toward a low-carbon society. Thus, Chap. 8 and the following chapters analyze the particular case of Nepal. Chapter 8 provides an overview of several aspects of climate change in Nepal such as the emission scenario, the climate change scenario, and impacts of climate change in five different sectors as well as poverty and opportunities for revenue generation from international climate change negotiations. Similarly, Chap. 9 focuses on the effect of climate trends on yields of basic food crops in Nepal. Chapter 10, the final chapter of the book, discusses the perception and realities of climate change in rural Nepal based on preliminary data generated through household surveys.

We think that this book, in addition to being a course material for graduate students, will fulfill the needs of the people seeking to understand the issues of climate change and its relation to agriculture and rural livelihoods in general and those of developing countries in particular, international climate change regimes related to agriculture and their impact on the economy of developing countries, including the opportunities resulting from mitigation and adaptation activities. It will also help the people to know the methodologies to asses the impact of climate change in agriculture, especially the impact on the issues of production, food prices and food insecurity. The empirical discussions on Nepal to grasp the issue in a local context, an addition to the dearth of such works, will be useful, even to the professionals, to enhance the understanding of the issue and its location specificity in the developing countries. We will value these and other readers and appreciate any comments and advices to improve the contents of the book.

We would like to acknowledge Luni Piya (Ph.D.) for her contributions to Chaps. 8–10. We would also like to thank Ph.D. students (IDEC, Hiroshima University) Suman Lal Shrestha and Mrinila Singh for their contributions to Chaps. 4 and 5, respectively. Last but not least, we are grateful to the Hiroshima International Center for Environmental Cooperation (HICEC), Graduate School for International Development and Cooperation, Hiroshima University, for providing the opportunity for this manuscript to be prepared for publication as course material for GELs students.

Hiroshima, Japan

Keshav Lall Maharjan
Niraj Prakash Joshi

Contents

Chapter 1
Background Information on Climate Change and Agriculture

Abstract This chapter intends to provide background information on climate change. It is done by providing a definition of climate change supplemented by some of the evidence suggested by the definition. We described the factors responsible for climate change, i.e., Greenhouse Gases (GHGs), their characteristics and sources, and changes in their concentration over time. In addition, brief description about how these GHGs warm earth surface through Greenhouse Gas effects is also provided. At the end of this chapter, relation of agriculture to climate is presented in the simplest terms.

Keywords Anthropogenic • Climate • Greenhouse Gases • Temperature

1.1 Climate Change

The climate system is a complex, interactive system consisting of the atmosphere, land surface, snow and ice, oceans and other bodies of water and living creatures. Basically, the atmospheric component of the climate system characterizes climate. Climate, often defined as average weather, is described in terms of the mean and variability of temperature, precipitation, and wind over a period of time, more specifically the classical period of 30 years (Le Treut et al. 2007).

United Nations Framework Convention on Climate Change (UNFCCC) (UN 1992, p. 3) defines climate change as "a change of climate which is attributed directly or indirectly to human activity that alters the composition of the global atmosphere and which is in addition to natural climate variability over comparable time periods." This definition is the narrow definition giving consideration only to the human activities. It is not only the human activities that alter the composition of the global atmosphere but it is also a natural variability itself. Therefore, the definition given by the Intergovernmental Panel on Climate Change (IPCC) in 2007 is accepted as the broader definition of climate change. IPCC (2007a, p. 30) defines climate change as

K.L. Maharjan and N.P. Joshi, *Climate Change, Agriculture and Rural Livelihoods in Developing Countries*, Advances in Asian Human-Environmental Research,
DOI 10.1007/978-4-431-54343-5_1, © Springer Japan 2013

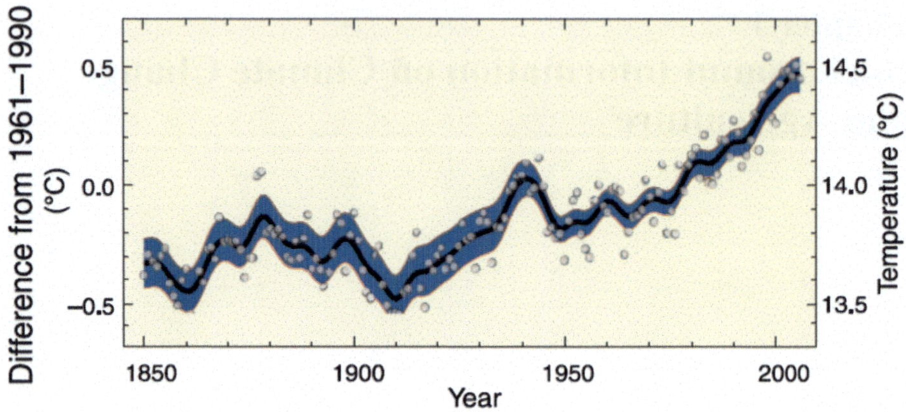

Fig. 1.1 Observed changes in global average surface temperature (IPCC 2007a). *Note*: Differences are relative to corresponding averages for the period 1961–1990. *Smoothed curves* represent decadal averaged values while *circles* show yearly values. The *shaded areas* are the uncertainty intervals estimated from a comprehensive analysis of known uncertainties and from the time series

"a change in the state of the climate that can be identified by changes in the mean and/or variability of its properties that persists for an extended period, typically decades or longer." Thus, Climate change refers to any change in climate over time, whether due to natural variability or as a result of human activity.

Evidences suggest a clear indication of climate change in the earth. Warming of the earth is unequivocal. Over the last few decades global average surface temperatures have been rising (Figs. 1.1 and 1.2). The global temperature reached its peak in 1998. Moreover, 11 of the last 12 years (between 1995 and 2006) are among the warmest years in the instrumental record of global surface temperature since 1850. Such increase in temperature has caused changes in weather patterns, widespread melting of snow and ice, and rising of global average sea level (IPCC 2007a).

There is a rise in the 100-year linear trend even between 1901–2000 and 1906–2005. The coefficient rose from 0.6 to 0.74 between these periods. Moreover, the linear warming trend over the later 50 years from 1956 to 2005 is nearly twice than that for the 100 years from 1906 to 2005. All these suggest that there is a higher rate of temperature rise in the recent years. This rise, however, is not likely to be uniform across the earth. The temperature increase will be greater in the higher latitudes and also at the night than during the day. This means there will be a decrease in the range of temperatures both through the day and across latitudes. Similarly, it is expected to warm at a larger rate during winter compared to the summer season.

In consistence with the temperature rise, there is an increase in sea level and decrease in snow and ice extent as well. The global average sea level rose at an average rate of 1.8 mm per year over the period 1961–2003. The rise is more intense if we consider the period between 1993 and 2003. Between these periods, sea level

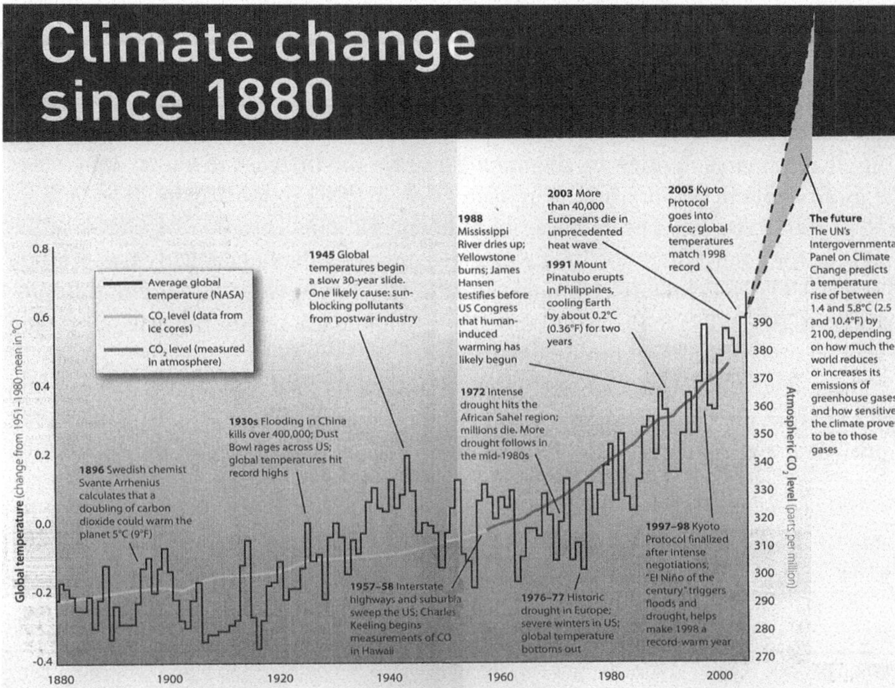

Fig. 1.2 Chronicles of climate change related indicators and responses at global scale (Henson 2006)

increased at the rate of 3.1 mm per year. Thermal expansion, decreases in glacier and ice caps, and losses from the polar ice sheets have contributed 57 %, 28 % and 15 %, respectively to the sea level rise since 1993 (IPCC 2007a).

There is annual shrinkage of the Arctic sea ice by 2.7 % per decade on average since 1978. The rate of decrease is larger in summer with an average loss of 7.4 % per decade. Similarly, mountain glaciers and snow covers also declined in both the hemispheres. In the Northern Hemisphere, seasonally frozen ground has decreased by about 7 % with decreased in spring of up to 15 % since 1900 (IPCC 2007a).

Besides, numerous long-term changes in weather patterns have also been observed. Over the period from 1900 to 2005, significant changes in precipitation trend are observed in many large regions. For instance, precipitation increased significantly in eastern parts of North and South America, northern Europe and northern and central Asia, whereas precipitation declined in the Sahel, the Mediterranean, southern Africa and parts of southern Asia. Moreover, it is more likely that the frequency of heavy precipitation events has increased over most areas causing water borne natural disaster in some areas and increased drought in the other. Similarly, there is an observational evidence of an increase in intense tropical cyclone activity together with a predicted increase in intense tropical cyclone activity.

1.2 Sources of Climate Change

Natural variability and human activities are the causes of climate change. The contribution of human activities to climate change, however, is increasing. The probability that human activities are the main cause for the increase in temperature since the mid-twentieth century has risen from 66 % in 2001 to more than 90 % in 2007 (IPCC 2001, 2007a). The increase in atmospheric concentration of GHGs is the major factor contributing to climate change caused by the human activities. Carbon dioxide (CO_2), methane (CH_4), nitrous oxide (N_2O), F-gases such as chlorofluorocarbons (CFC_{11}), hydro fluorocarbons (HFC_{23}) and carbon tetra fluoride (CF_4), and water vapors are the most prominent GHGs.

The concentration of CO_2 has risen significantly from 280 parts per million (ppm) before the pre-industrial era to 379 ppm in 2005 (Table 1.1). This is mainly contributed by increased use of fossil fuel (Fig. 1.3) in transportation, building,

Table 1.1 Concentration of Greenhouse Gases from pre-industrial to current time (IPCC 2007b; Blasing 2010)

GHGs	Before 1750	1998	2005	2010	Atmospheric lifetime (years)
Carbon dioxide (CO_2)	280 ± 20 ppm	365 ppm	379 ppm	386 ppm	50–200
Methane (CH_4)	~700 ppb	1,745 ppb	1,774 ppb	1,866 ppb	12
Nitrous oxide (N_2O)	~ 270 ppb	314 ppb	319 ppb	323 ppb	114

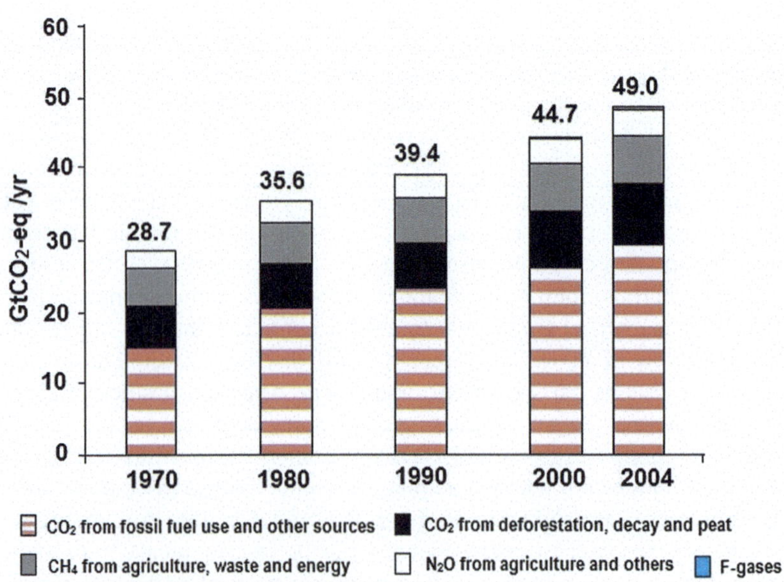

Fig. 1.3 Type of Greenhouse Gases from various sources (IPCC 2007a)

heating and cooling, and the manufacture of cement and other goods. Plants and ocean soak up a huge amount of CO_2, which check the level of CO_2 from increasing rapidly. However, deforestation has resulted in an imbalance of such natural process of the CO_2 sink (Fig. 1.3). Besides decay of plant matter and respiration process of both human and animal also releases CO_2 as part of natural processes.

Concentrations of CH_4 and N_2O have also risen significantly (Table 1.1). It has risen from 770 parts per billion (ppb) to 1,745 ppb and 270 ppb to 314 ppb for CH_4 and N_2O, respectively. CH_4 emission is the outcome of human activities related to agriculture, natural gas distribution and landfills. Emissions from paddy field, ruminant livestock, as well as improper management of animal excreta are the major agricultural activities contributing to increasing concentration of CH_4. A natural process that occurs in wetlands is also contributing to increased concentration of CH_4. CH_4 is perceived as a powerhouse GHG as it absorbs 21 times more infrared energy within its atmospheric lifetime of 12 years compared to what CO_2 does over roughly a century.

Use of external input such as fertilizer and fossil fuel burning contributes to increased concentration of N_2O in the atmosphere. Natural processes in soils and the ocean also release N_2O. The total CO_2-eq of these prominent GHGs is estimated to be around 455 ppm CO_2-eq in 2005. Unless the concentration is stabilized below 550 CO_2-eq, a harmful irreversible consequence of climate change through a temperature rise of more than 2 °C is inevitable (IPCC 2007a).

1.3 Greenhouse Gas Effects

The idea of greenhouse effects emerged from the evidence that although the sun's light and heat easily pass through glass and other transparent materials, heat from other non-transparent sources does not. Thus, solar radiation is the main component that powers the climate system. There are three fundamental ways that change the radiation balance of the earth (Fig. 1.4), which are as follows:

1. By changing the incoming solar radiation
2. By changing the fraction of solar radiation that is reflected, which is called "albedo" and is caused by the changes in cloud cover, atmospheric particles or vegetation, and
3. By altering the long-wave radiation from the earth back towards space, which is mainly caused by changes in Greenhouse Gas concentration

The long term balance between the amounts of incoming solar radiation absorbed by the earth and the atmosphere is maintained by the earth and the atmosphere by releasing the same amount of outgoing long-wave radiation. The surface absorbed around 47 % of the incoming solar radiation (Trenberth et al. 2009). This energy is transferred to the atmosphere by warming the air in contact with the surface, by evapo-transpiration and by long-wave radiation that is absorbed by clouds and Greenhouse Gases. The atmosphere in turn radiates long-wave energy back to earth as well as out to space.

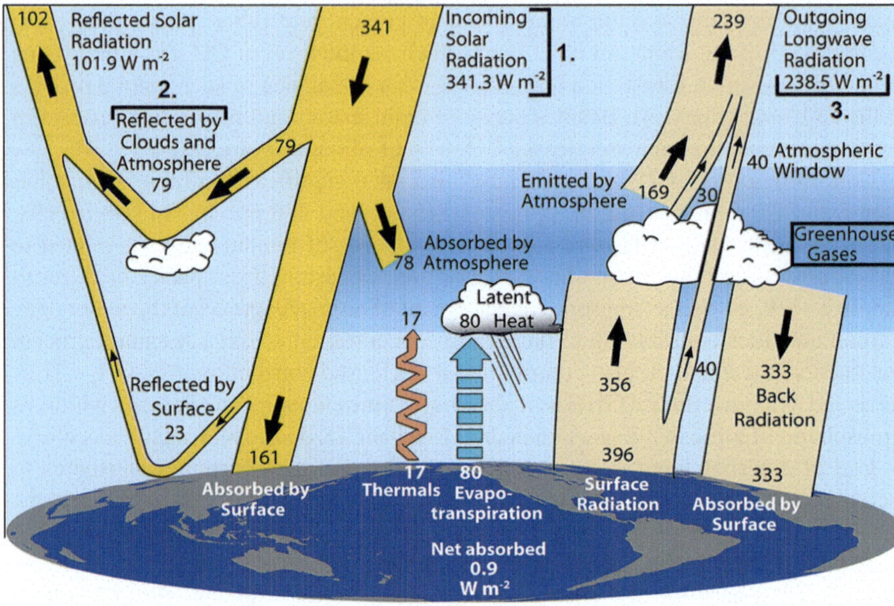

Fig. 1.4 Estimate of the earth's annual and global mean energy balance (Trenberth et al. 2009)

Box 1.1 Greenhouse Gas Effect

The Sun powers Earth's climate, radiating energy at very short wavelengths, predominately in the visible or near-visible (e.g., ultraviolet) part of the spectrum. Roughly one-third of the solar energy that reaches the top of Earth's atmosphere is reflected directly back to space. The remaining two-third is absorbed by the surface and, to a lesser extent, by the atmosphere. To balance the absorbed incoming energy, the Earth must, on average, radiate the same amount of energy back to space. Because the Earth is much colder than the Sun, it radiates at much longer wavelengths, primarily in the infrared part of the spectrum (see Fig. 1.4). Much of this thermal radiation emitted by the land and ocean is absorbed by the atmosphere, including clouds, and reradiated back to Earth. This is called the greenhouse effect. The glass walls in a greenhouse reduce airflow and increase the temperature of the air inside. Analogously, but through a different physical process, the Earth's greenhouse effect warms the surface of the planet. Without the natural greenhouse effect, the average temperature at Earth's surface would be below the freezing point of water. Thus, Earth's natural greenhouse effect makes life possible in the Earth. However, human activities, primarily the burning of fossil fuels and clearing of forests, have greatly intensified the natural greenhouse effect, causing global warming.

Source: Le Treut et al. (2007).

The ability to generate an artificial warming of the Earth's surface was demonstrated in a simple greenhouse experiment to provide analogy of the greenhouse effect. It recognized that the air itself could also trap thermal radiation. Joseph Fourier in 1824, argued that "the temperature [of the earth] can be augmented by the interposition of the atmosphere, because heat in the state of light finds less resistance in penetrating the air, than in repassing into the air when converted into non-luminous heat." Later, John Tyndall (1861) identified that the thermal radiation is absorbed by complex molecules (cited from IPCC 2007b). He noted that changes in the amount of any of the radiatively active constituents of the atmosphere such as H_2O or CO_2 could have produced "all the mutations of climate which the researches of geologists reveal." In the 1970s some other radiatively active gases (Greenhouse Gases) such as CH_4, N_2O and CFCs were widely recognized as the important anthropogenic Greenhouse Gases. The increased concentration of these gases has resulted in an increase in proportion of long-wave radiation bounced back to the earth's surface from the atmosphere. The proportion has increased from 83 % on average of 5 years during the mid-1980s to 84 % on average during 2000 to 2004 (Kiehl and Trenberth 1997; Trenberth et al. 2009). In a nutshell, other remaining the same, the more Greenhouse Gases there is the less radiation can escape from the earth to the space, and the warmer we get.

1.4 GHGs and Global Warming

As we discussed in the earlier section, GHGs like CO_2, CH_2, and N_2O etc. has a greenhouse effect on the earth. This means that the higher concentration of GHGs in the atmosphere will result in the warming of the earth. However, there is also a chance that the heating and cooling itself could cause changes in GHGs concentration in the atmosphere. For instance, when global temperatures become warmer CO_2 is released from the oceans. Thus, increasing CO_2 concentration may amplify the warming by enhancing the greenhouse effect. Inversely, with cooling of the earth CO_2 enters the ocean and contributes to additional cooling through reduced CO_2 concentration. Figure 1.5 shows the long term (650,000 years) changes in CO_2 level and temperature. During this period there have been seven major climate shift; approximately one about every 100,000 years. Both warming and cooling phases have taken place even in the absence of human activities. But it is evident that throughout the period of more than 650,000 years there is a strong correlation between temperatures and CO_2. Though this correlation does not imply causation, a high degree of correlation provides safe logic to assume that one variable likely affects the other, directly or indirectly. Over these long natural cycles, CO_2 increases have lagged temperature increases at the beginning of a cycle, but then the increasing CO_2 content in the atmosphere caused further temperature increase.

The further increase in temperature due to the increasing CO_2 content can be established by the more recent relation of CO_2 concentration and global temperature (Fig. 1.6). This is further indicated by both the basic physics of the greenhouse

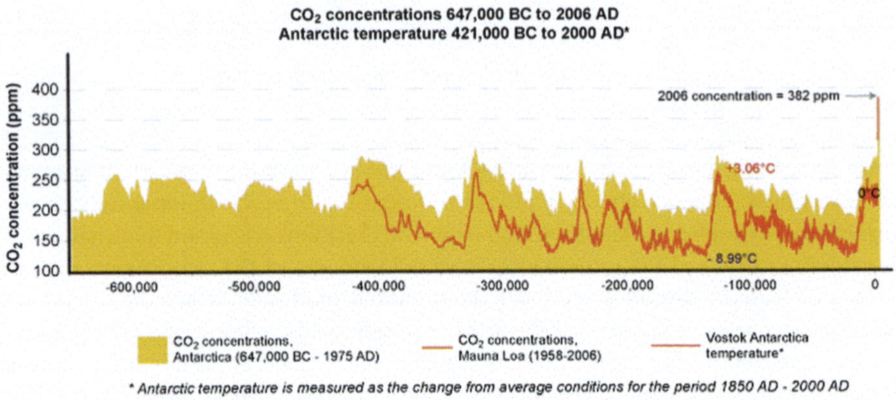

Fig. 1.5 Fluctuations in temperature (*solid line*) and the atmospheric concentration of CO_2 (*shaded area*) over the past 649,000 years (EPA 2010). *Note*: The *vertical bar* at the end is the increase in atmospheric CO_2 levels over the past two centuries and before 2007

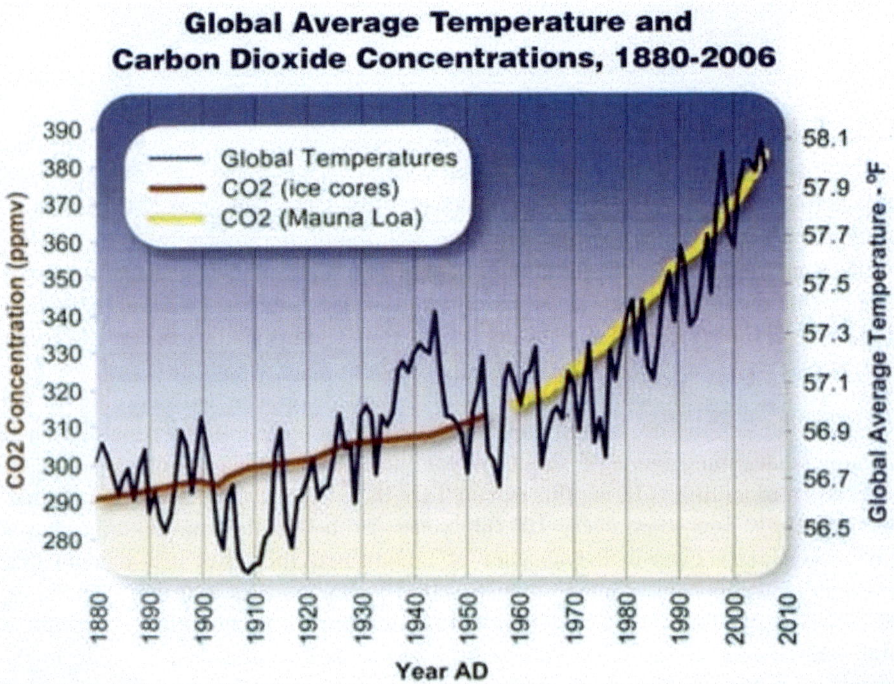

Fig. 1.6 Global average temperature and carbon dioxide concentrations, 1880–2006 (Cherry and Braasch 2008)

effect of CO_2 and other GHGs, and more detailed calculations using sophisticated models of atmospheric radiative transfer (Ward 2010). The figure suggests that the earth's surface temperature has risen over the past 100 years. It has increased 0.74±0.18 °C during the twentieth century. Such level of increase in temperature is not seen in at least several hundred years and even much longer. This increase has mostly been attributed to the increasing concentrations of GHGs. Even if atmospheric concentrations of GHG had been held steady at their 2000 levels warming of the earth is inevitable. Under such scenario a further warming for the next two decades at a rate of about 0.1 °C per decade is predicted, which is mainly due to the slow response of the oceans in absorbing atmospheric CO_2 (IPCC 2007a). IPCC (2007a) predicts the increase in temperature between 2 and 4.5 °C under a doubling of CO_2 scenario. All these suggest the strong positive correlation between atmospheric GHGs and temperature

1.5 Climate and Agriculture

Agriculture is defined as the science, art, and business of cultivating the soil, producing crops and raising livestock. In the broadest sense, agriculture comprises the entire range of technologies associated with the production of useful products from plants, and animals through soil cultivation, and crop and livestock management. It also implies the activities of processing and marketing of such products. Thus, the primary agricultural products consist of crop plants for human consumption as well as animal feed and also the livestock products.

The climate is one of the most important factors influencing agriculture. Climate is defined as the weather averaged over a long period of time, the standard averaging period of 30 years, of a certain region. It basically encompasses the statistics of temperature, rainfall, wind, humidity, atmospheric pressures, atmospheric particle count and other meteorological elements. The climate of a location is affected by its latitude, terrain, and altitude. Climate can be classified according to the average and the typical range of the climate variables, most commonly temperature and rainfall.

Crop studies have revealed that crops and livestock are highly sensitive to interannual variations in climate. Year-to-year fluctuations in temperature and precipitation lead to large annual losses for farmers across the world (Mendelsohn 2000). The effect of warming on agriculture in developing countries is uncertain because these countries use more labor intensive methods and they are located in lower latitudes. The share of agriculture in anthropogenic emissions is around 15 %, which exceed 32 % if we consider land use change. Hence it is important to study climate change and agriculture relationship. Therefore, emissions of different sectors worldwide as well as the climate change–agriculture relationship will be dealt in more details in Chap. 3 and thereafter, whereas Chap. 2 focus on GHG emissions.

References

Blasing TJ (2010) Recent Greenhouse Gas concentrations. Retrieved 2 February 2011, from Carbon Dioxide Information Analysis Center: http://cdiac.ornl.gov/pns/current_ghg.html

Cherry L, Braasch G (2008) How we know what we know about our changing climate: scientists and kids explore global warming. Dawn Publications, Nevada City

EPA (2010) Climate change-science: past climate change. Retrieved 16 February 2011, from U.S. Environmental Protection Agency: http://epa.gov/climatechange/science/pastcc.html

Henson R (2006) The rough guide to climate change: the symptoms, the science and the solutions. Rough Guides Limited, London

IPCC (2001) Climate change 2001: the scientific basis. Cambridge University Press, Cambridge

IPCC (2007a) Climate change 2007: synthesis report. Working group contribution to the fourth assessment report of Intergovernmental Panel on Climate Change. Intergovernmental Panel on Climate Change (IPCC), Geneva

IPCC (2007b) Climate change 2007: mitigation of climate change. Cambridge University Press, Cambridge

Kiehl J, Trenberth KE (1997) Earth's annual global mean energy budget. Bull Am Meteorol Soc 78(2):197–206

Le Treut H, Somerville R, Cubash U, Ding Y, Mauritzen C, Mokssit A et al (2007) Historical overview of climate change. In: Soloman S, Qin D, Manning M, Chen Z, Marquis M, Averyt KB et al (eds) Climate change 2007: the physical science basis. contribution of working group I to the fourth assessment report of the intergovernmental panel on climate change. Cambridge University Press, Cambridge, pp 94–127

Mendelsohn R (2000) Measuring the effect of climate change on developing country agriculture. In: Two essays on climate change and agriculture: A developing country perspective (FAO Economic and Social Development Paper 145). Food and Agriculture Organization of the United Nations, Rome, pp 2–31

Trenberth KE, Fasulla JT, Kiehl J (2009) Earth's global energy budget. Bull Am Meteorol Soc 90(3):311–323

UN (1992) United Nations framework convention on climate change. UNFCCC/INFORMAL/84. United Nations (UN), New York

Ward RE (2010) Comment: 'knock, knock: where is the evidence for dangerous human-caused global warming?' by Robert M Carter. Econ Anal Policy 40(2):125–143

Chapter 2
Global Scenario of Greenhouse Gas Emission

Abstract Emission of GHGs has a huge significance on global climate change regardless of where has it been emitted. Therefore, having knowledge on from where did GHGs come from, and what are the main sources is very important in order to share the responsibility to cope with climate change. Therefore, this chapter provides the information on historical trends of emissions for major emitting countries and the contribution by different sectors on total emissions over time.

Keywords Carbon equivalent • Carbon flux • Carbon reserve • Developing countries • Industrialized countries

2.1 Amount of GHGs Stored in the Earth

Carbon is held in the Earth mainly by the atmosphere, the terrestrial biosphere, the ocean and the sediments. Freshwater systems and non-living organic materials are examples of terrestrial biosphere that hold carbon. In case of the ocean, carbon is stored in the form of dissolved inorganic carbon as well as living and non-living marine biota. Fossil fuels and coal deposits are some of the important forms of the sediment that store carbon. At present, the earth holds around 46,913 Gt[1] carbon (Table 2.1).

The ocean is the largest store of carbon primarily in the form of CO_2. It stores more than 85 % of total global carbon (Table 2.1). But the ocean carbon does not rapidly exchange it with the atmosphere. Towards the pole, seawater in the ocean becomes cooler that facilitates CO_2 to be easily dissolved in the water, thus, favoring the uptake of CO_2 from the atmosphere. However, as it moves towards the equatorial belt, ocean surface release CO_2 due to warm temperatures. This indicates that with warming of sea water CO_2 is not easily absorbed by the ocean, and remains in the atmosphere. The absorption and release of atmospheric carbon in the ocean can also

[1] One giga ton is equal to one billion metric ton.

K.L. Maharjan and N.P. Joshi, *Climate Change, Agriculture and Rural Livelihoods in Developing Countries*, Advances in Asian Human-Environmental Research, DOI 10.1007/978-4-431-54343-5_2,

Table 2.1 Carbon storage in different earth system (Riebeek 2001)

Earth system	Storage area	Giga tons (Gt)
Atmosphere	Atmosphere	750 (1.6)
Terrestrial biosphere	Vegetation	610 (1.3)
	Soils	1,580 (3.37)
Ocean	Surface ocean	1,020 (2.17)
	Deep ocean	38,100 (81.21)
	Marine biota	3 (0.01)
	Underwater dissolved organic carbon	~700 (1.49)
	Ocean sediments	150 (0.32)
Sediments	Fossil fuels and cement production	4,000 (8.53)
Total		**~46,913**

Note: Figures in parentheses indicate percentage

take place through thermo-haline circulation in the ocean. For instance, cold, downward moving currents such as those that occur over the North Atlantic absorb CO_2 and transfer it to the deep ocean, whereas, upward moving currents such as those in the tropics brings CO_2 up from depth and release it to the atmosphere.

The atmosphere holds 800 Gt of carbon, which is a rise from 750 Gt (Riebeek 2001; U.S. DOE 2008). Carbon in the atmosphere is primarily held in the form of CO_2, CH_4 and CFCs. Carbon dioxide is the most prominent form of carbon. As of 2002, humans were adding close to 26 Gt of CO_2 in the atmosphere every year. Burning of fossil fuels and cement production has a significant share in this addition. Of the total addition about 50 % of CO_2 is absorbed by the earth system, while the rest 50 % remains in the atmosphere. The major concern for the present days is the steep climbing of the global emission rate of CO_2. The rate has sharply increased from 15 Gt in 1970 to its present value of close to 26 Gt. This rate primarily depends on the rate of economic growth. The rate can level off or even decline slightly during the worldwide economic recession/slowdown, as it happened in 1992 and 1993. Though the emissions have reduced during the recession, the total amount of CO_2 (CO_2 concentration atmosphere in ppm) in the air, however, has risen every year since measurement (Table 1.1). We can see the sharp increase in the concentration of three important GHGs, i.e., CO_2, CH_4, and N_2O after nineteenth century (Fig. 2.1). CO_2 equivalent (see Box 2.1 for details) concentration level had reached 420 ppm by the middle of 2008 as the emissions continue to rise at the rate of 3 % every year (Sparatt 2009).

Vegetation and soils are major components of terrestrial biosphere. This system is neutral in terms of adding carbon in the atmosphere. Rather, terrestrial biosphere serves as a sink of atmospheric carbon (Fig. 2.2). However, if terrestrial biosphere is not properly managed it could become a major source of carbon to the atmosphere as the emission rate of the terrestrial biosphere system surpasses the sink rate. It is becoming a very crucial issue being debated in recent climate change related negotiations. The process of sedimentations requires significantly longer time horizon to capture carbon. For instance, it will take millions of years to form fossil fuels, which is an important form of sediments. However, due to excessive extractions that far surpass the rate of its formation, sediments contribute to emissions of carbon in the atmosphere. The latest figure illustrates that 9 Gt of carbon is released into the atmosphere from the sediments in the form of fossil fuel burning and cement manufacturing (Fig. 2.2).

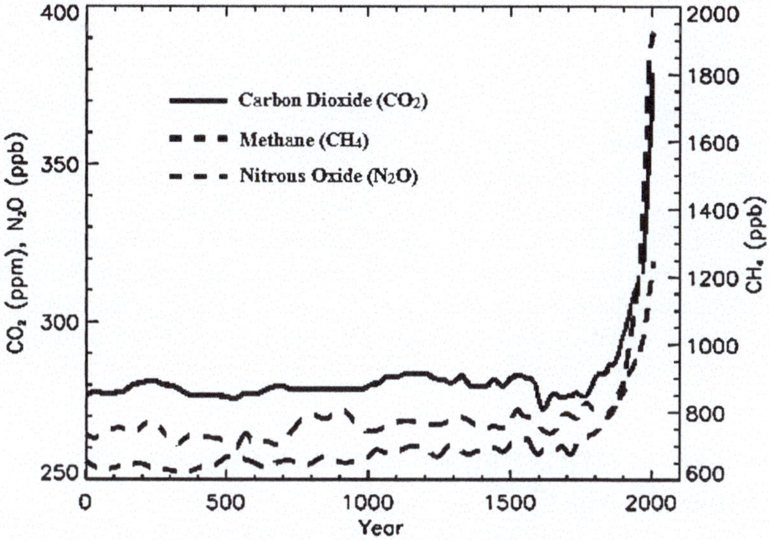

Fig. 2.1 Concentration of greenhouse gases from 0 to 2000 in the atmosphere (Le Treut et al. 2007)

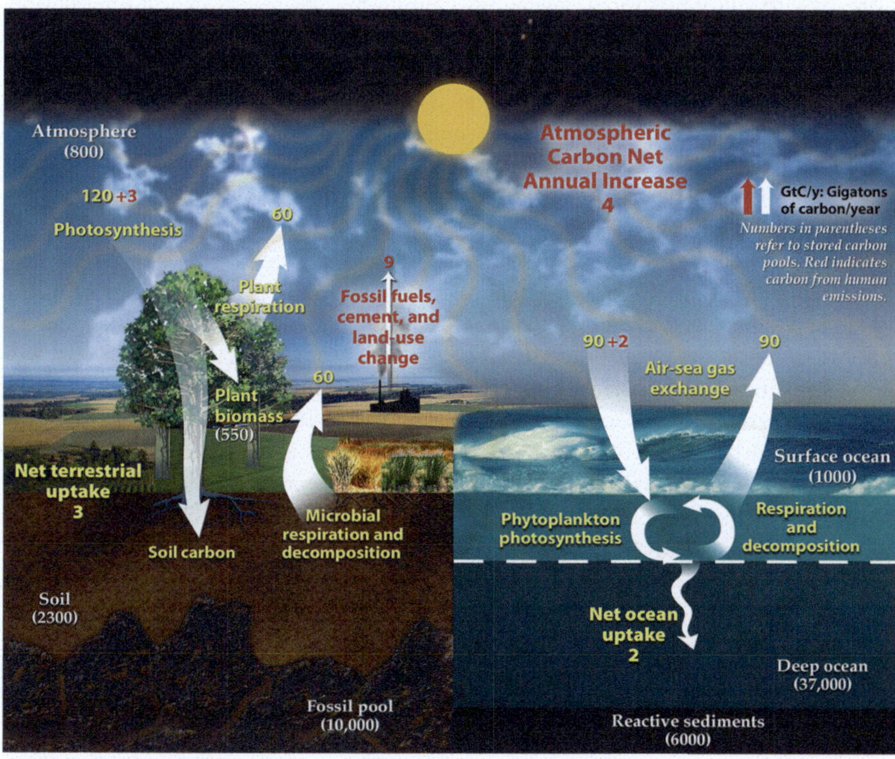

Fig. 2.2 Carbon reserve and annual carbon flux (U.S. DOE 2008)

Box 2.1 CO_2 Equivalent

Because of variation in the power of the GHGs climatic effects, researchers often have to rely on a unit called global warming potential (GWP). This is measure of both its greenhouse potency and lifespan in the atmosphere as compared to those of CO_2. Methane (CH_4) for example, is shorter-lived than CO_2 but much more powerful in its ability to trap heat in the atmosphere. Therefore, a GWP of CH_4 is estimated to be somewhere around 23 compared to value of 1 for CO_2. Similarly, a GWP for N_2O is estimated to be around 310. These figures can be multiplied by the prevalence of each gas to produce a carbon equivalent that enables all emission to be considered in a single standard unit.

2.2 Source of GHGs Emissions: Which Country Emits the Most?

As we are all aware GHGs remains in the atmosphere with consequent effects on climate change, it is equally important to study the historical emissions together with the emission being produced today. A share of cumulative emissions over time is one way to study historical emissions. Cumulative emissions over a long period provide an indication of the country's total contribution to GHG concentrations in the atmosphere within the timeframe considered. The historical share in cumulative emissions measured over the period from 1900 to 2005 places the US at the top, contributing 30 % of the cumulative CO_2 emissions within the period (Fig. 2.3). The US is followed by the European Union, China, Japan, and India contributes 23 %, 8 %, 4 % and 2 %, respectively in historical emissions of CO_2, whereas the rest of the world contributed only 33 % (IEA 2007).

Despite the very nominal contribution of developing countries (like China and India) in historical emission, the gap between the industrialized (like US, European Union, and Japan) and developing countries (like China and India) is in decreasing trend. Estimation made by International Energy Agency (IEA) (IEA 2007) shows that in the reference scenario China's share of cumulative emissions from 1900 to 2030 rises to 16 %, approaching closer to the US (25 %) and the European Union (18 %). Similarly, India's cumulative emissions will be same as that of Japan (4 %). The same study on the assumption of high growth scenario, shows that cumulative emissions of China will be same as that of the European Union by 2030, whereas India will surpass Japan (IEA 2007).

The closing gap between the industrialized and developing countries in terms of cumulative emission of GHGs can be more clearly understood by studying the recent trends of emissions.

Figure 2.4 shows the dynamics of CO_2 emissions of the top emitters in recent years. The emission of CO_2 from these 25 countries account more than 80 % of total global CO_2 emissions. The US had the biggest share to CO_2 in the global

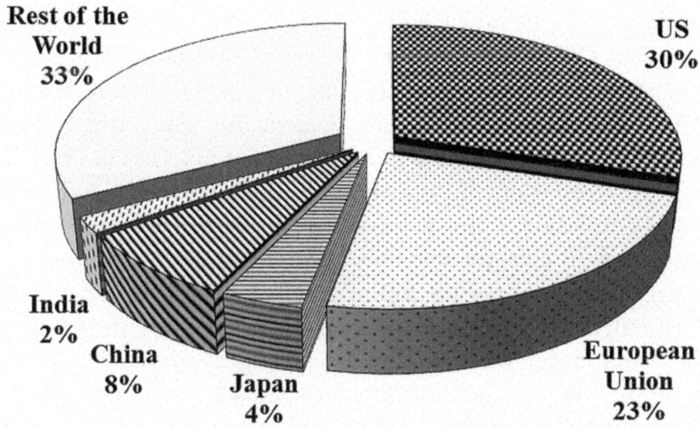

Fig. 2.3 Cumulative emission of energy related CO_2 by region, 1900–2005 (IEA 2007)

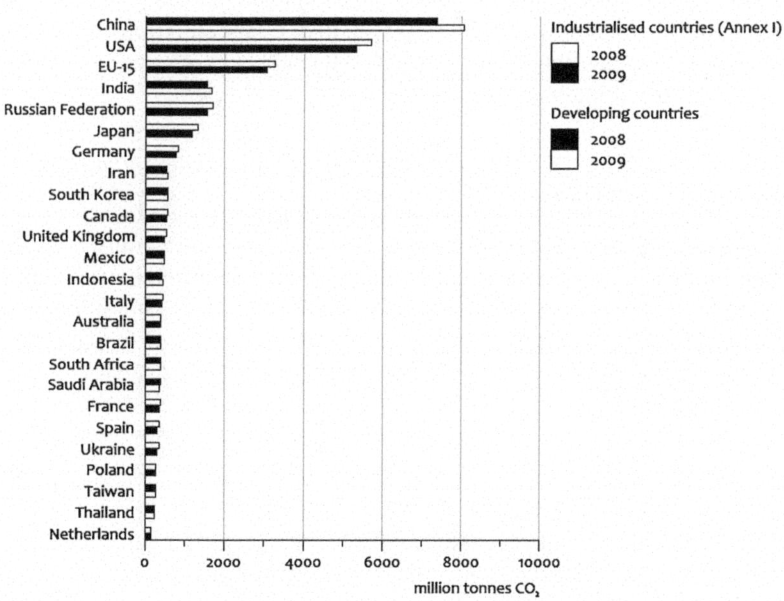

Fig. 2.4 Topmost CO_2 emitting countries in 2008 and 2009 (Olivier and Peters 2010)

anthropogenic CO_2 emission till 2006. However, since 2007 China took the number one position and emitting the largest proportion of CO_2. Not only China but also many other developing countries has increased global share of annual CO_2 emission. Even the projection shows that the emission from developing countries, especially China and India, will grow continuously (Fig. 2.4).

It is observed that developed/industrialized as well as developing countries are putting successful effort to curb their CO_2 emission and are able to cut their emissions. Reduction in the emission is also attributed to the global recession. The similar decrease in CO_2 emissions was observed in 1974–1975, 1980–1982 and 1992 due to recessions caused by a large increase in oil price (Olivier and Peters 2010). All the industrialized countries, except for Australia, Poland and the Netherlands, fall under the top most CO_2 emitter are able to cut emissions in 2009 compared to 2008. Even some of the developing countries like Saudi Arabia are able to cut their CO_2 emissions. However, some developed countries like Australia, Poland, and the Netherlands are able only to stabilize their emission. Such stability is achieved also by many of the top emitters falling under the developing countries like Mexico, Brazil, South Africa, Taiwan, and Thailand. But, the emission in China, India, Iran, and Indonesia was an increasing trend even in 2009 and expected to grow further (Fig. 2.4). An increase in growth of CO_2 emissions has occurred in these countries regardless of the credit crunch that affected most industrialized/developed countries and a large compensation by the industrialized countries that have emission mitigation targets under the Kyoto Protocol (Olivier and Peters 2010).

It is difficult to establish which country is responsible for climate change simply by totaling up the amount of annual or cumulative GHG emissions in the country. We also have to consider the country's total GHG emission relative to population or economy, i.e., carbon intensity (Box 2.2).

Box 2.2 Carbon Intensity

Carbon intensity is a measure of how much carbon a country emits to produce certain amount of economic output. Thus, carbon intensity is the emission pro-rated by the Gross Domestic Product (GDP) and gives emission per unit of GDP. Similarly, it can be measured in terms of population as well, which gives per capita emission.

List of top 25 countries having the highest per capita emission is shown in Fig. 2.5. This figure shows that Australia has the highest per capita emission, which is also in increasing trend. We can observe the increasing trend in Spain and Canada as well. In contrast to this, there is decreasing trend of per capita emission for almost all other developed/industrialized countries. In case of developing countries, however, there is a sharp rise in per capita emission. The stark increase in per capita emission can be observed in many developing countries, especially in China, Iran, Taiwan, and Saudi Arabia. The increase in emission is persistent in all developing countries.

Emission per unit of GDP reflects the energy efficiency of the economy. In this measure of GHG emissions, developing countries take the top positions. Seven out of the ten highest emitters per unit of GDP are from developing countries. Higher emissions per unit of GDP reflects the lower efficiency of the economy in terms of GHG emissions in relation to GDP. This means developing countries emit more GHG to produce the same amount of GDP. But we can also observe the sharp decline in emissions per unit of GDP in these developing countries. This implies

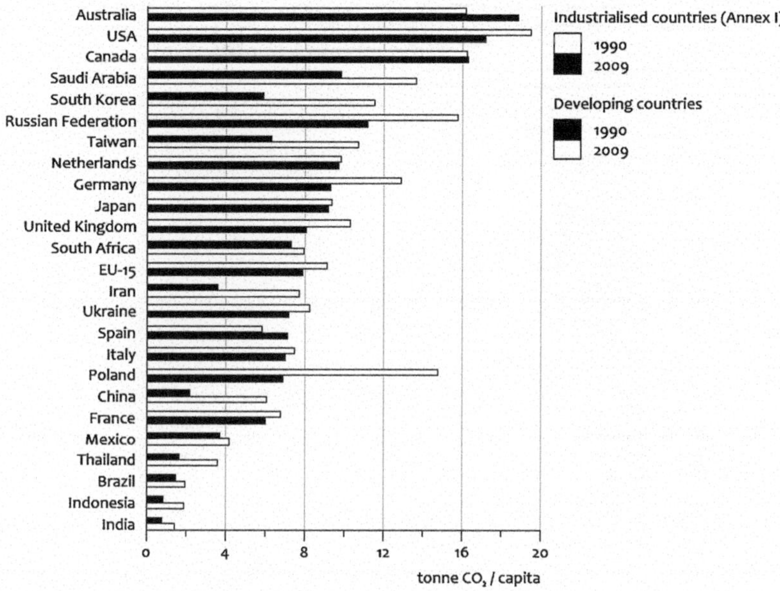

Fig. 2.5 Countries having highest CO_2 emission per capita in 1990 and 2009 (Olivier and Peters 2010)

that developing countries like China, South Africa, India, Taiwan and Mexico are able to achieve a reduction in GHG emissions per unit of GDP, i.e., achieved energy efficiency in 2009 compared to 1990 (Fig. 2.6). Brazil is not able to gain the energy efficiency, but is able only to stabilize the energy efficiency. There are several other developing countries which are not able to improve their energy efficiency. For instance, in countries like Iran, Saudi Arabia, Indonesia and Thailand, energy efficiency is deteriorating and is emitting more GHGs in 2009 than in 1990, to produce the same amount of GDP.

Evidences on GHG emissions provided so far suggest that though the cumulative share of developing countries is less compared to that of developed/industrialized countries, the share is increasing. In some case like China, the share surpasses that of developed/industrialized countries and even contributes the biggest share of total global emission since 2006. It is due to the more rapid population and GDP growth and their increasing share of energy-intensive industries. However, emission per capita in developing countries is still far below than that of the developed/industrialized countries. But developed countries are able to reduce the per capita emission during the period of 1990 and 2009, whereas it is ever increasing in the case of developing countries. This signifies that the energy intensity is increasing in developing countries compared to the developed countries due to rapid population growth and economic growth in the developing countries fueled by conventional energy sources i.e., fossil fuels especially coal, and petroleum products. Though the energy intensity is increasing in many developing countries, some of the leading

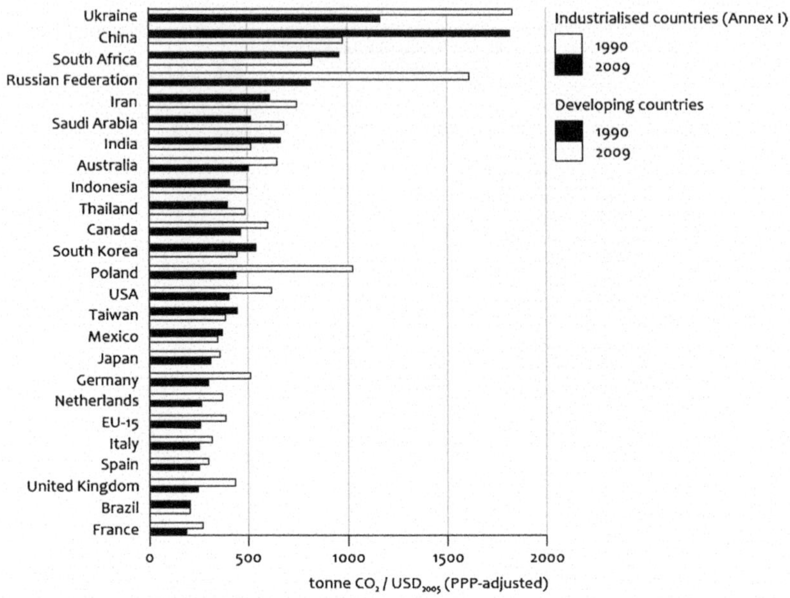

Fig. 2.6 Countries having highest CO_2 emission per unit of GDP in 1990 and 2009 (Olivier and Peters 2010)

developing countries like China, South Africa, and India are also able to increase energy efficiency resulting in reductions of GHG emissions per unit of GDP between 1990 and 2009. Thus, the concern should not always be focused on curbing the emission in developing countries, which share in global emission is ever increasing, but the focus should also be given to checking the high growth rate of emission in developing countries through justifiable international regimes that compensate them for their marginal share in historical emission.

2.3 Source of GHG Emissions: Which Sector Emits the Most?

Energy is the most important sector in terms of emitting the significant proportion of total global GHG emissions. It comprises electricity and heat, manufacturing and construction, transportation, other fuel combustion and fugitive emissions. All of these sub-sectors emit more than 64 % of global GHG emissions. There is a continuous increase in the amount of GHG emissions from these sub-sectors between 1990 and 2005 (Fig. 2.7). Most of the emission from energy is from coal followed by petroleum (Fig. 2.8). Coal is the most inefficient source of energy in terms of emission but due to its abundance supply, it remains the cheapest source of energy for many developed as well as developing countries. Natural gas is another

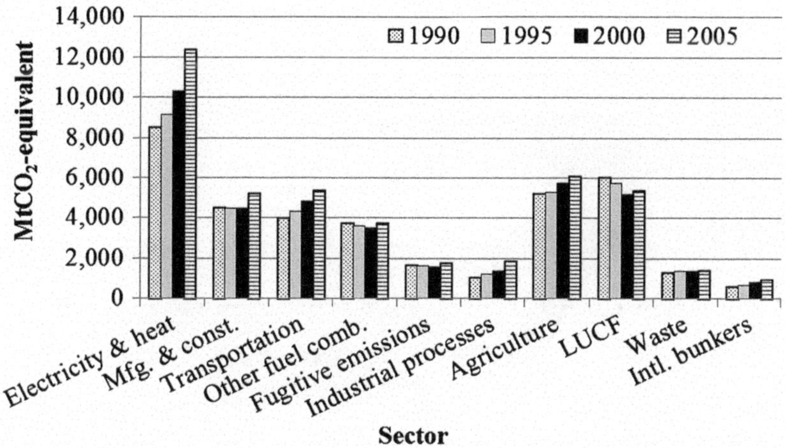

Fig. 2.7 Global GHG emissions by sector, 1990–2005 (WRI 2011). *Note*: Mfg. & const. is manufacturing and construction, other fuel comb. is other fuel combustion, LUCF is Land-Use Change and Forestry, and Intl. bunkers is International bunkers

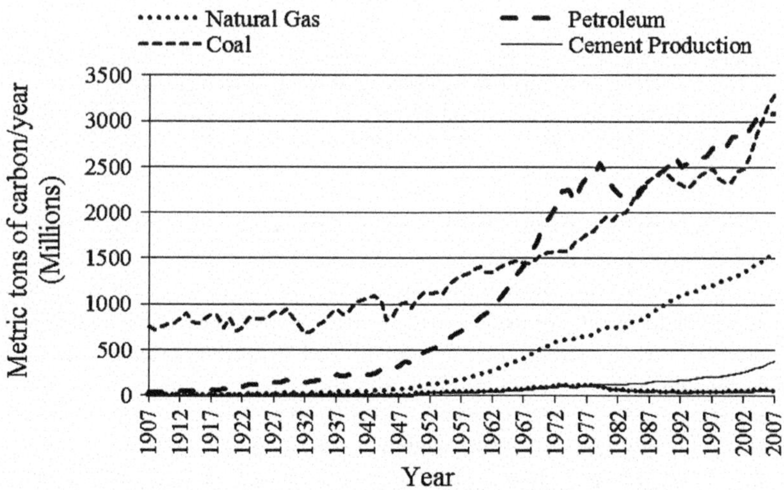

Fig. 2.8 Emission scenarios from various sources of energy (Boden et al. 2010)

important source of energy, which is considered to be an efficient source of energy. However, its share in total GHG has been increased mainly due to the increased use of natural gas as the source of energy.

The minute details on the emissions from each of the sub-sectors can be seen in Fig. 2.9. GHG emissions from energy sector are growing at the annual rate of 1.6 % between these periods. The rise in GHGs from energy sectors can be attributed to the rapid growth of the emissions from developing countries. Within the energy

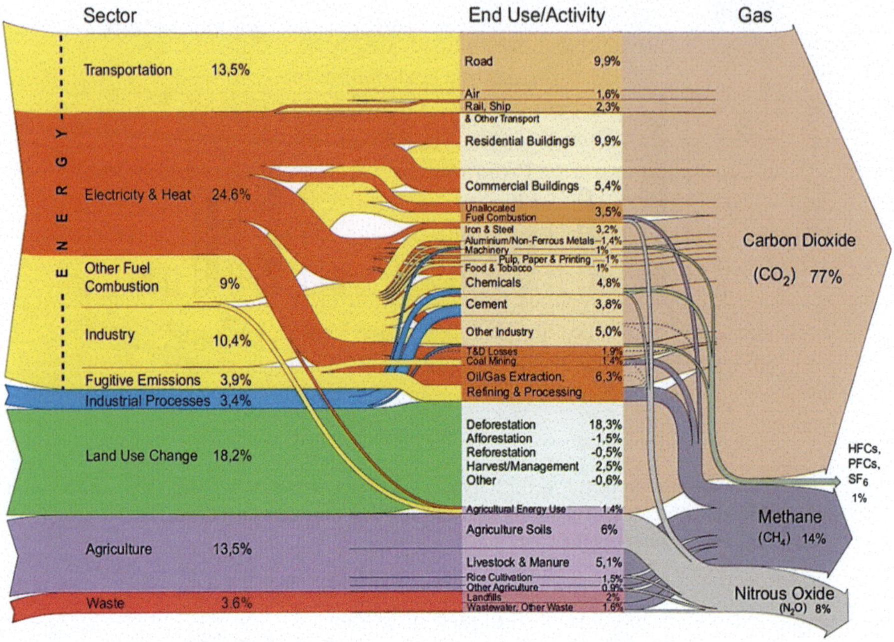

Fig. 2.9 World greenhouse gas emissions by sector (Baumert et al. 2005). *Note*: All the data is for 2000. All calculations are based on CO_2-eq, using 100-year global warming potentials from the IPCC, based on a total global estimate of 41,755 $MtCO_2$-eq. Land use change includes both emission and absorptions. *Dotted lines* represent flows of less than 0.1 % of total GHG emissions

sector, GHG emissions from electricity and heat, and transportation sub-sectors show the highest annual growth rate.

The annual growth rate of the emissions from electricity and heat is 2.5 % (Stern 2006). The growth of the emissions from electricity and heat is the highest between 1990 and 2005, which increased by around 45 %. Developing countries, particularly China, India, and other Asian countries and the Middle East, were among those, which experienced the greatest growth in emissions over this period. Fossil fuels account for three quarters of the fuel used in electricity and heat sub-sectors, of which coal is the most dominant and also responsible for the majority of emissions from this sector. Therefore, choice of fuel also affects the emission from electricity and heat. Coal is the most inefficient fuel, which is followed by oil and gas. The replacement of coal by oil, gas, or development of technology would use all the sources efficiently which would help reduce the growth rate of emission to a greater extent.

The emissions from transportation are another important sub-sector of energy, which was growing at the rate of 2 % per year between 1990 and 2005. Three-quarters of transportation emissions are from road transport, and one-quarter is shared by aviation, and rail and shipping equally. As the emissions from the transportation sector expanded by 35 % between 1990 and 2005, it is the second-fastest growing sector after electricity and heat. The emissions from transportation are

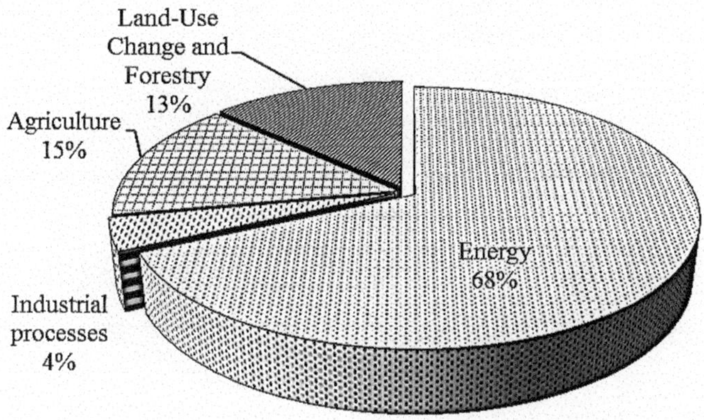

Fig. 2.10 Share of different sectors on global GHG emissions, 2005 (WRI 2011)

expected to be more than doubled by 2050. Therefore, transportation is considered as the sector where the most trouble appears to be emerging especially in the developing world where transport-related emissions are growing at a high rate.

Income is the key driver for such rapid growth in emission from the electricity and heat as well as transportation. Demand for electricity and heat increases with the increase in income. Thus, the countries expected to experience the fastest growth in the emissions are those which are expected to have a higher economic growth. About half of the increase in the emission from these sectors between now and 2030 could be from growth in India, China, and Africa (Stern 2006). Similarly, with an increase in income of people they want to travel using more carbon-intensive modes. For instance, as people get richer they switch from public vehicle to private vehicle. However, the emission from transportation is influenced by the cost. An increase in prices of fuel or private vehicle itself tends to choke-off relatively less demand in the transport sector than it does in the building and industry sector.

Industrial process accounts 4 % of global GHG emissions in 2005 (Fig. 2.10). This is a significant jump from 2.9 % in 1990. The emission from this sector is growing at the fastest rate with the growth rate of 4 % between 1990 and 2005. Achievements in attaining efficiency in energy by the world's most technologically advanced nations have resulted in a decline in emissions from industrial process in such developed countries since 1990. However, this achievement is counterbalanced by the explosion of industry in the world's growing economies, such as China and India.

Agriculture was responsible for 15 % of global greenhouse gas emissions (that excludes waste and international bunkers from Fig. 2.7) in 2005 (Fig. 2.10). Within agriculture sector, fertilizers are the largest single source of emissions. It accounts 38 % of the emissions from agriculture. Fertilizers increase the emission of N_2O from the natural processes of nitrification and denitrification. Livestock is the second largest source of emission, which accounts 31 % of the emissions from agriculture. CH_4 is the major GHG emitted by livestock, which is produced as a waste product of digestion by ruminants, particularly cattle. This process is known as enteric

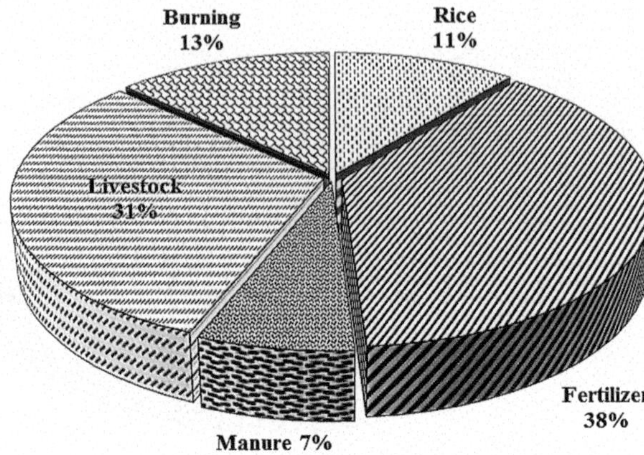

Fig. 2.11 Sources of non-CO_2 GHG emissions from the agriculture sector, 2000 (Stern 2006)

fermentation. Burning of savannah and agricultural residues, and open burning from forest clearing contributes 13 % of GHG emissions from agriculture (Fig. 2.11).

Wetland rice cultivation is another important source of GHG emissions in agriculture accounting 11 % of agricultural emissions. Anaerobic decomposition of organic matter produces CH_4 in the flooded rice fields. Therefore, the level of emissions from rice cultivation is dependent upon the specific water management practices and quantity of organic matter involved. Manure management methods also contribute to CH_4 emissions. It causes 7 % of agricultural emissions. Methane is emitted when the manure is not stored in a sufficiently oxygenated environment that leads to anaerobic decomposition. Similarly, as nitrogen in livestock manure and urine encourages nitrification and denitrification, N_2O is released due to such anaerobic decomposition.

The agriculture sector is also associated with the CO_2 emission via soil and biomass management practices that disturb the natural carbon sinks. Besides, the agriculture sector is also indirectly responsible for emissions from other sectors. For instance, agriculture is a key driver for land use change such as deforestation. Similarly, use of agricultural equipment that requires an energy source and transportation of agricultural inputs and output lead to emissions from the industry, energy and transport sectors, respectively. The majority of agricultural emissions emerge from developing countries. In the case of CH_4 emission from wetland rice cultivation, around 90 % of emissions currently come from China and South East Asia (Stern 2006).

In the period from 2000 to 2020, the emissions from agriculture are expected to rise almost 30 %. It is expected that almost 67 % of this increase is expected to come from developing countries. This is supported by the fact that in recent years developing countries have accounted for an increasingly large share of agricultural emissions. The share has increased from 67 % in 1990 to 75 % in 2000 and is projected to reach 80 % in 2020. If we consider the source, around 50 % of the projected growth in emissions is expected to come from the use of fertilizer on agricultural soils (Stern 2006). In the case of agriculture as well income and population growth are the key drivers behind the growth in the emissions from this sector.

Land-Use Change and Forestry (LUCF) account 13 % of global GHG emissions. Deforestation, which is highly concentrated in a few countries, is the main driving force for emission from LUCF. For instance, around 30 % and 20 % of land-use GHG emissions are from Indonesia and Brazil, respectively. Conversion of land from forest to agriculture, as demand for agriculture is driven by population and income, is the primary driver for land-use changes. Similarly, demand for forest product is another important driver of land use change emissions. South East Asia is a good example where intensive logging is fuelled by the strong demand for timber from fast growing regional economies. The emissions from this sector is projected to fall by 2050 as deforestation is assumed to saturate after 85 % of the forest has been cleared (Stern 2006). In addition, with proper policies and enforcement mechanisms in place, the rate of deforestation could be checked and substantial amount of GHGs emissions cut can be achieved.

References

Baumert KA, Herzog T, Pershing J (2005) Navigating the number: Greenhouse gas data and international climate policy. World Research Institute, Washington

Boden T, Marland G, Andres R (2010) Global, regional, and national annual CO_2 emissions from fossil fuel burning, cement manufacture and gas flaring: 1751–2007. Carbon Dioxide Information Analysis Center, Tennessee, USA. Retrieved 14 February 2011, from Carbon Dioxide Information Analysis Center: http://cdiac.ornl.gov/ftp/ndp030/CSV-FILES/

IEA (2007) World energy outlook 2007: China and India insights. Organization of Economic Cooperation and Development (OECD) and International Energy Agency (IEA), Paris

Le Treut H, Somerville R, Cubash U, Ding Y, Mauritzen C, Mokssit A, Prather aM (2007) Historical overview of climate change. In: Soloman S, Qin D, Manning M, Chen Z, Marquis M, Averyt KB, Miller MT (eds) Climate change 2007: the physical science basis. Contribution of Working Group I to the Fourth Assessment Report of the Intergovernmental Panel on Climate Change. Cambridge University Press, Cambridge, pp 94–127

Olivier JG, Peters JA (2010) No growth in the total global CO_2 emissions in 2009. Netherlands Environmental Assessment Agency (PBL), Bilthoven

Riebeek H (2001) The carbon cycle. Retrieved 1 May 2011, from NASA Earth Observatory: http://earthobservatory.nasa.gov/Features/CarbonCycle/carbon_cycle2001.pdf

Sparatt S (2009) Assessing the alternatives: financing climate change mitigation and adaptation in developing countries. Stamp Out of Poverty, London

Stern N (2006) The economics of climate change: the Stern review. Cambridge University Press, Cambridge

U.S. DOE (2008) Carbon cycling and biosequestration: integrating biology and climate through systems science. Report from the March 2008 Workshop, DOE/SC-108. U.S. Department of Energy Office of Science (U.S. DOE), Washington, DC

WRI (2011) Climate analysis indicators tool (CAIT) version 8.0. Retrieved 10 February 2011, from World Resource Institute (WRI): http://cait.wri.org/cait.php

Chapter 3
Effects of Climate Change on Plant and Animal Physiology

Abstract Agriculture is highly dependent on weather or climate, it has very strong relation with climate change. Therefore, this sector is more dependent upon and vulnerable to climatic conditions than any other human activity. Thus, a possible effect of global climate change on agriculture received an important place in the public's perception. While both natural and managed ecosystems are projected to experience direct consequences of global climate change, it is the managed ecosystems of food production that became an immediate pressing concern for researchers as well as policy makers as it is directly associated with billions of human lives. This chapter, therefore, will highlight how the various indicators of climate change have an impact on agriculture. In doing so, this chapter firstly put a brief note on plant physiology and then describes how the plant physiology is being affected by climate change mainly through soil characteristic, and disease and pest infestation.

Keywords Agriculture • Disease • Pest • Soil characteristics • Temperature

3.1 Plant Physiology

Optimum amounts of various inputs such as sunlight, water, CO_2, nutrients and limited weeds, diseases and insects are determining factors for crops to grow and be economically productive. In the process of plant growth and development, which is also a very complex process, photosynthesis converts the energy of sunlight into chemical potential energy stored in the organic structures of plants. Putting it simply, the plant uses sunlight to convert CO_2, water, nitrogen and other resources to Oxygen (O_2), water and carbohydrates. This process occurs only in the presence of sunlight i.e., during the daytime. Thus, sunlight, CO_2 (absorbed through pores in the plant), water (irrigation), nitrogen and other resources in the form of fertilizers serve as input in the process resulting in production of O_2, water, and carbohydrate. The carbohydrate is used by plants for its growth and development, whereas O_2 is released

K.L. Maharjan and N.P. Joshi, *Climate Change, Agriculture and Rural Livelihoods in Developing Countries*, Advances in Asian Human-Environmental Research,
DOI 10.1007/978-4-431-54343-5_3,

into the atmosphere and water becomes a part of already present water in the body of plants. The relative distribution of dry matter among the plant's organs determines the efficiency of the growth process relative to economic yield (Morison 1996). The process of photosynthesis also has a relation with temperature in terms of effectiveness of the process as a whole and also through increased demand for water.

3.2 Climate Change and Plant Physiology

As we discussed in the earlier section of this book, climate change is mainly associated with atmospheric CO_2 concentration and the key weather variables such as rainfall, temperature, and solar radiation as well as other climate extremities like flood, drought, landslide and etc.

Carbon dioxide being one of the important inputs in the photosynthesis process, its atmospheric concentration is an important factor affecting plant growth and yield. Its effect on plant growth and yield is both direct as well as indirect. For instance, increase in photosynthesis and growth is a direct effect of CO_2 rise, whereas reduction in plant water loss is an indirect effect of CO_2 rise. The elevated level of CO_2 concentration leads to reduced stomata openness resulting in reduced water loss. Thus, increase in atmospheric concentration of CO_2 will result in increased photosynthesis rate and also increased stomatal resistance in crops leading to overall increased water-use efficiency. Due to this beneficial relation of increasing level of atmospheric CO_2, these processes have been coined as "CO_2 fertilization" (Fig. 3.1).

Temperature and humidity will also play an important role in agricultural production. Increase in temperature would prolong the crop growing seasons, by shortening the growing period of plants, in areas where they are now limited by cold temperatures especially at high latitudes and high elevations. Similarly, as a warmer atmosphere can hold more water vapor, increase in precipitation will have beneficial effects on crops in some semi-arid locations. However, the area and extent of any such regions of enhanced precipitation is not precisely known.

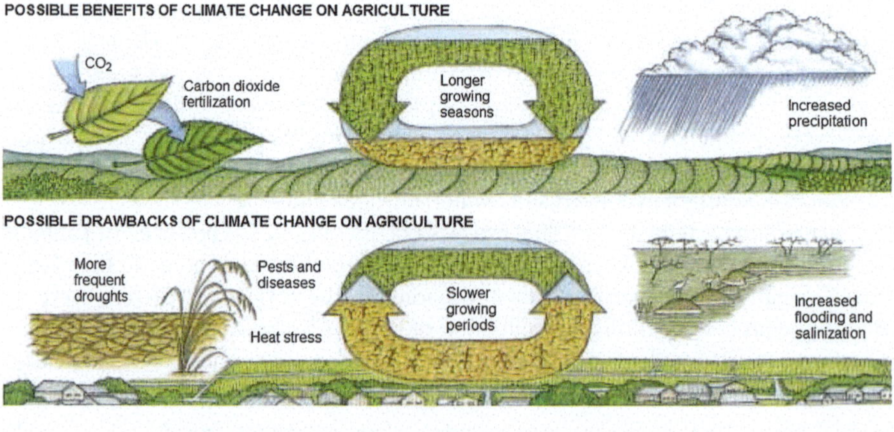

Fig. 3.1 Possible effects of climate change on agriculture (Bongaarts 1994)

It is not always true that the climate change brings the positive effect on agriculture. Climate change will also have adverse effects on agriculture. The development rate of plants is dependent on combinations of temperature and CO_2 to a larger extent. Elevated CO_2 and temperature levels increase the rate of the plant development thereby shortening the growing period, which may result in less potential growth. Increased temperature speeds up crops physiologies through their growing cycle especially in the grain filling stage. Since total yield is a product of the rate and duration of grain filling, shortening of grain filling duration due to higher temperature tends to exert a negative pressure on the yield of most annual crops (Bongaarts 1994; Rosenzweig and Hillel 2005).

Moreover, the effects of elevated CO_2 depend on nature of crops whether it follows C4 photosynthetic pathway or C3 (see Box 3.1 for details). Those crops with the C4 photosynthetic pathway are already optimized at the current CO_2 level, therefore, have only a small response to higher CO_2. On the other hand, crops with the C3 photosynthetic pathway will respond the most to rising CO_2 level. Even within the C3 plants increasing temperature and carbon dioxide could be beneficial for the production of crops such as lettuce or spinach; the yield of which is defined as total above ground dry matter. In contrast, there is no any synergistic effect between CO_2 and temperature when it comes to the crops, yield of which is measured as reproductive yield (seed). In some cases like that of rice, sorghum and dry bean, even a negative effect is observed between yield with elevated CO_2 and temperature (Mendelsohn and Dinar 2009).

The increased evaporation from soil and plants due to increased temperature must coincide with more rainfall in a region in order to avoid more frequent dry spell and droughts, which are detrimental to many of the agricultural crops. Similarly, a rise in temperature, mostly in tropical and sub-tropical areas, will reduce crop yield (Bongaarts 1994). This is because, in these areas certain crops are already grown near their limit of heat tolerance, and further increases in temperature will result into heat stress. Agricultural regions close to the ocean are highly vulnerable due to possible sea level rise and associated saltwater intrusion. In addition, flooding can harm crops through impeded soil aeration and salinization. Adverse impact on agriculture due to climate change is further aggravated by the increased incidence of pest and diseases favored by the frequent drought, heat stress, and fast growing period as well as increased flooding.

The discussion provided in the earlier paragraphs shows that climate change has beneficial as well as harmful consequences on agriculture in the world. However, some of these possible beneficial aspects of climate change on agriculture are yet to be validated under field condition. Moreover, possible benefits of climate change are expected to concentrate on high latitude and high altitude regions. Though the overall responses of agriculture to climate change is expected to be positive, in the long run it is expected to shift from positive to neutral, and then to negative even in high latitude and high altitude regions (Rosenzweig and Hillel 2005). Thus, in the long run climate change will have adverse impact on agriculture at the global level in the absence of technological advancement. This adverse effect will be skewed more towards developing countries, which lies mostly in low latitude regions. Furthermore, developing countries lack resources, which is very critical in devising appropriate adaptation measures to meet changing agriculture condition. Therefore, the combination of potentially greater climate stresses and low adaptive capacity in developing countries expose them to a higher degree of vulnerability to climate change.

Box 3.1 Difference Between C3 and C4 Plants/Crops

Based on the number of carbon compound produced during the initial process of photosynthesis, plant species can be divided into C3 plants or C4 plants. Accordingly, crop species are also known as C3 crops and C4 crops.

C3 plants/crops follow the photosynthetic pathways as follows:

- Add carbon dioxide (CO_2) from atmosphere to a phosphorylated 5-carbon sugar called Ribulose Bisphosphate
- Reaction catalyzed by the enzyme Ribulose Bisphosphate Carboxylase Oxygenase (RUBISCO) will takes place
- This results into 6-carbon compound breaks down into two molecules of 3-phosphoglyceric acid (PGA) (Here initially two different compounds are produced namely; 3-phophoglycerate and 3-phophoglycolate. Since 3-phosphoclycolate cannot be used in the photosynthetic carbon reduction cycle, it must be recycled to phosphoglycerate via the photorespiratory pathway. Therefore, due to additional requirement of CO_2 for the photosynthesis process, it is C3 plant/crops that benefits through carbon fertilization from increased atmospheric CO_2 concentration.)
- These 3-carbon compounds serve as the starting material for the synthesis of glucose and other food molecules. Thus, the pathway is called the C_3 pathway, and the process is called the Calvin cycle.

C4 plants/crops follow the photosynthetic pathways as follows:

- The CO_2 is inserted into a 3-carbon compound (C_3) called Phosphoenolpyruvic Acid forming the 4-carbon compound oxaloacetic acid (4-carbon compounds).

C3 plants account majority of earth's plant species. These plants flourish in cool, wet, and cloudy climates, where light levels may be low, because the metabolic pathway is more energy efficient, and if water is plentiful, the stomata can stay open and let in more carbon dioxide. However, carbon losses through photorespiration are high. As the increased CO_2 tends to suppress photo-respiration, higher levels of atmospheric CO_2 would stimulate photosynthesis in the C3 plants. Thus, this category of crops species will be more responsive to carbon fertilization. C3 plants include most of the crops species such as wheat, rice, barley, cassava, and potato.

C4 plants, which inhabit hot, dry environments, have very high water-use efficiency, so that there can be up to twice as much photosynthesis per gram of water as in C3 plants, but C4 metabolism is inefficient in shady or cool environments. C4 plants include such tropical crops as maize, sugar cane, sorghum and millet, which are important for the food security of many developing countries, as well as pasture and forage grasses.

Source: Furbank and Taylor (1995); UNEP and UNFCCC (2002).

3.3 Physical Factors Affecting Plant Physiology

This section deals with the some physical factors that affects agricultural production. The first part discusses how soil quality will be affected by several aspects of climate change like increased atmospheric CO_2 concentration, unpredictable rain, increased temperature, and sea level rise. Soil quality being the vital factor of production for agriculture any adverse impact on it will negatively affect agricultural production. Discussion will then be followed by the possibility of insect pest infestation thereby agricultural production due to change in climate.

3.3.1 Soil Characteristics

Soil characteristics/properties are crucial determinant of agricultural production. Some of its properties are transitory and vary rapidly such as nutrient content, whereas some are virtually permanent taking fairly long time to alter such as texture. Changes in temperature, rainfall or CO_2 are some of the factors associated with climate change that affect transitory properties of soil. For, instance change in climate will affect soils in a terrain by affecting their erosion potential with rainfall, vegetation cover and cultivation; and changes in atmospheric CO_2 concentration determine changes in soil organic matter quantity and type (Brinkman and Sombroek 1996; Adams et al. 1998).

Thus, the global changes in climate will directly affect the soil-forming factors like organic matter supply from biomass, soil temperature regime and soil hydrology. In simple words, moisture level and nutrient availability in the soil are the most important edaphic factors to be influenced by or to influence the climate change. Projected change in climate is expected to have an adverse impact on soil quality due to the influence on several soil processes that affect moisture level and nutrient availability in the soil (Brinkman and Sombroek 1996). These factors are related to crop yield, thereby crop production.

With the projected change in climate during the next century the following soil-forming factors (forcing variables) are expected to occur (Brinkman and Sombroek 1996):

- A gradual and continuing rise in atmospheric CO_2 concentration increases organic matter supplies to soils due to CO_2 fertilization, i.e., increased photosynthetic rates and water-use efficiencies.
- As a result of changes in air temperature and vegetation zones, there will be minor increases in soil temperatures in the tropics and subtropics, and moderate increases and extended periods in which soils are warm enough for microbial activity in temperate and cold climate.
- Increase in temperature that leads to extension of growing period lead to minor increases in evapotranspiration in the tropics to major increases in high latitudes.
- Peak rainfall intensities could increase in several regions. For instance, there will be an increase in amount and in the variability of rainfall in the tropics, possible

decrease in rainfall in a band in the subtropics pole-ward of the present deserts, and minor increases in amount and variability in temperate and cold regions.
- A gradual sea-level rise. This will cause deeper and longer inundation in the river and estuary basins and on levee back-slopes. Consequently, there will be brackish-water inundation that leads to encroachment of vegetation accumulating pyrite in soils near the coast.

These expected changes in soil-forming factors (factor variables) can lead to the processes with an attendant adverse impact on soil quality, important among such processes are as follows (Brinkman and Sombroek 1996):

- Hydrolysis: Water containing CO_2 removes silica and basic cations from the soil.
- Cheluviation: Dissolves and removes especially Al and Fe by chelating organic acids.
- Ferrolysis (a cyclic process of clay transformation and dissolution mediated by alternating iron reduction and oxidation): Decreases the cation exchange capacity by aluminium interlayering in swelling clay minerals.
- Dissolution (of clay minerals by strong mineral acids): produces acid aluminum salts and amorphous silica.
- Reverse weathering: clay formation and transformation under neutral to strongly alkaline conditions, which may create compounds like montmorillonite, palygorskite or analcime.

Increase in temperature may accelerate hydrolysis and cheluviation due to increased leaching rates (Brinkman and Sombroek 1996; Lal 2005). Similarly, Ferrolysis occurs where soils are subject to reduction and leaching in alternation with oxidation. With the warming of the world, this may happen over larger areas than at present especially in high latitudes and in monsoon climates. Dissolution of clay minerals by strong acids is halted by an improvement in drainage. This means oxidization of sulphidic materials in coastal plains is enhanced through drainage improvement. However, the likelihood of occurrence of natural drainage would be reduced by a rise in sea level thereby increasing dissolution. Reverse weathering would continue in most of the presently arid areas and is expected to start in areas drying out during global warming. All these processes increase soil erodibility, and decrease water and nutrient retention capacity.

With all this, there exist two schools of thought with regard to the effects of projected climate change on soil quality. The first school of thought argues that climate change is likely to increase risk of soil degradation. There are ample evidences of the accelerated soil erosion and other degrading processes, which affected soil quality especially in developing countries of the tropics and subtropics (Table 3.1). Deforestation and agricultural mismanagement are the most significant causative factors of soil degradation in Asia. But in Africa it is overgrazing and over exploitation of vegetative cover that are causing soil degradation (Oldeman et al. 1991). Thus, developing Asia and Africa share significant proportion of global soil erosion from all means such as water erosion, wind erosion, chemical degradation as well as physical degradation. Expected climate change is expected to accelerate soil degradation especially in ecologically sensitive regions such as the Himalayan-Tibetan

Table 3.1 Soil degradation in selected regions (million hectares) (Oldeman et al. 1991)

Region	Soil erosion by water	Soil erosion by wind	Chemical degradation	Physical degradation
Africa	227.4	186.5	61.5	18.7
Asia	440.6	222.2	73.2	12.1
South America	123.2	41.9	70.3	7.9
Central America	46.3	4.6	6.9	5
World total	1,093.7	548.3	239.1	83.3

Note: Chemical degradation involves loss of soil nutrients, soil salinization, urban-industrial pollution and acidification. Physical degradation involves compaction, waterlogging and subsidence of organic soils

ecosystems, the un-terraced slopes of China and Southeast Asia, tropical areas of Southeast Nigeria, the semi-arid Sahelian region of West Africa, sloping lands of Central America and the Andean valleys and cerrado region of South America (Scherr and Yadav 1996).

Regional variation in climate change will have different effects on different regions. For instance, increased risks of soil erosion by water are projected in the regions expected to have higher rainfall intensity (Nearing et al. 2004; Easterling et al. 2007), whereas the regions expected to become drier will be susceptible and become more severely affected by wind erosion (Wiliams et al. 2002). Thus, as a consequence of increased temperature, total erosion is predicted to increase in coming days.

Salinization is another important aspect of soil quality, which is expected to be aggravated by increased temperature. There will be a higher risk of salinization in arid and semi-arid regions mainly due to increased water loss below the crop root zone (van Ittersum et al. 2003; Easterling et al. 2007). The Indus, Nile, Tigris, and Euphrates river valleys; northeast China; northern Mexico; and the Andean highlands are some of the region most prone to salinization (Scherr and Yadav 1996). Around 20 % of total irrigated area is estimated to be already affected by salinization due to which 12 million hectares of irrigated land may have already gone out of production (Nelson and Mareida 2001). Thus, there is a higher chance that there will be an exacerbation of the desertification problem especially in Sub-Saharan Africa (UNEP 1997).

In a nutshell, soil degradation mainly through accelerated erosions involves depletion of organic matter in soil. Therefore, soil organic carbon pool in most of the degraded soils is below their potential set by ecological factors. Increased temperature also accelerates losses of soil organic carbon. Depletion of soil organic carbon exacerbates nutrient depletion in low-input agricultural systems that are already vulnerable to severe nutrient depletion. Thus, sustainability of agriculture has been already problematic in areas like Sub-Saharan Africa, Sahel, and South Asia (Stoorvogel and Smaling 1990; Reardon 1995; Rhodes 1995; Easterling et al. 2007) including the Indus, Nile, Tigris, and Euphrates river valleys; northeast China; northern Mexico; and the Andean highlands (Scherr and Yadav 1996).

The second school of thought claims that soil quality may be improved in some ecosystems by the projected changes in climate. For instance, the CO_2 fertilization would increase biomass productivity with more litter and crop residues with a higher C:N ratio returned to the soil. This will result to decrease in the rate of mineralization and reduction in the turnover rate. Consequently, there will be a gradual increase in soil fertility due to increase in soil organic matter (Allard et al. 2004; Easterling et al. 2007). Besides, an increase in biological nitrogen fixation can also be attributed to such increase in soil fertility (Brinkman and Sombroek 1996). In addition, fungal "infection" of plant roots increases mycorrhizal activity, which enhance phosphorus (P) uptake (Lal 2005). All these positive developments lead to increase in root growth, which enhance the soil organic carbon pool and also the activity of soil biota such as earthworms and termites thereby improve soil structure. Thus, there will be decreased rate of soil erosion as well as reduction in leaching losses of plant nutrients. The detail of how the increase in atmospheric CO_2 concentration will improve soil quality is presented in Fig. 3.2.

Similarly, an increase in temperature enhances soil microbial activity, which accelerates the weathering of parent material with a resulting increase in the rate of new soil formation, and thus in agroecosystem soil loss tolerance. Also the enhanced microbial activity would strengthen elemental recycling mechanisms, increase the soil organic carbon pool, and improve the soil structure. Therefore, the first school of thought can be contended, and said that the projected climate change may improve soil quality and increase soil resilience.

Box 3.2 Effects of Higher Atmospheric CO_2, Unpredicted Rainfall, Increased Temperature and Sea Level Rise in Soil Quality

Higher atmospheric CO_2: Higher CO_2 will increase the growth rates and water-use efficiency of crops and natural vegetation, given other factors is not limiting. Therefore, there will be an improvement in soil quality thereby agricultural productivity accompanied by more crop residues, a greater total root mass and exudation, increased mycorrhizal colonization and activity of other rhizosphere or soil micro-organisms including symbiotic and root-zone nitrogen fixers.

Change in rainfall pattern: Increase in rainfall would increase leaching rates in well-drained soils with high infiltration rates and may cause temporary flooding with the resultant loss of fertile top-soil and also reduce organic matter decomposition. This would give rise to amount and frequency of run-off on soils more particularly in sloping terrain with sedimentation downslope, especially through the mudflows and/or landslides. Moreover, it would lead to soil salinization in low-lying areas or areas with the high ground water table. However, if the soils are most resilient against such changes, there would be adequate cation exchange and anion sorption, which minimize nutrient loss during leaching flows, and have a high structural stability and a strongly heterogeneous system of continuous macro-pores to maximize infiltration and rapid bypass flow through the soil during high-intensity rainfall.

(continued)

Box 3.2 (continued)

In case of decreased rainfall, there will be less dry-matter production, which lowers soil organic matter content.

Increased temperature: Higher temperature means higher evaporative demand, therefore demands more water. Similarly, increase in temperature leads to higher respiration rates, shorter periods of seed formation, and lower biomass production, thereby lessens soil organic carbon. Moreover, increased temperature may directly lower crop yields as well as grain quality, as it result in a shorter grain filling period, thereby smaller and lighter grains. However, at the same time increase in temperature would further stimulate microbial activity. Also, increased CO_2 would tend to counteract adverse effects of increased temperature. For instance, the water demand will be lessened under higher atmospheric CO_2 as it increases the potential evapotranspiration and higher water-use efficiency.

Sea level rise: A rise in sea level would tend to erode and move back existing coastline. There are also high chances that the area of perennially or seasonally saline soils would extend due to the penetration into further inland than at present by tidal flooding carrying saline water. Similarly, higher storm surges associate with a rising sea level could destroy existing coastal eustatic peat swamps, which might eventually be replaced by saltwater lagoons.

Source: Brinkman and Sombroek (1996); Adams et al. (1998); Easterling et al. (2007).

However, there is still some skepticism on the second school of thought as there is possibility that under the low nutrient inputs the total soil carbon sink may saturate at elevated CO_2 concentration (Gill et al. 2002; Easterling et al. 2007). Similarly, uncertainty still remains with respects to several key issues like stability of carbon and soil organic matter pool as a result of increased frequency of extremes (Easterling et al. 2007). For instance, there was significant soil carbon loss due to the recent European heat-wave in 2003 (Ciais et al. 2005). Similarly, the plant function may indirectly affect carbon storage as the effects of air pollution. As the ozone has the negative effect on biomass productivity and changes, the predicted increase in tropospheric ozone may result in significantly less enhancement of CO_2 sequestration rates even under elevated CO_2 (Loya et al. 2003). Therefore, in order to capture any possible benefits of climate change by any particular ecosystems, the adaptive options and/or the use of recommended management practices are the crucial determinants (Lal 2005).

3.3.2 Pest Infestation

Pest infestation here refers to possible invasion by insects or weeds or diseases or their combination that harm the plant physiological process thereby reduces the

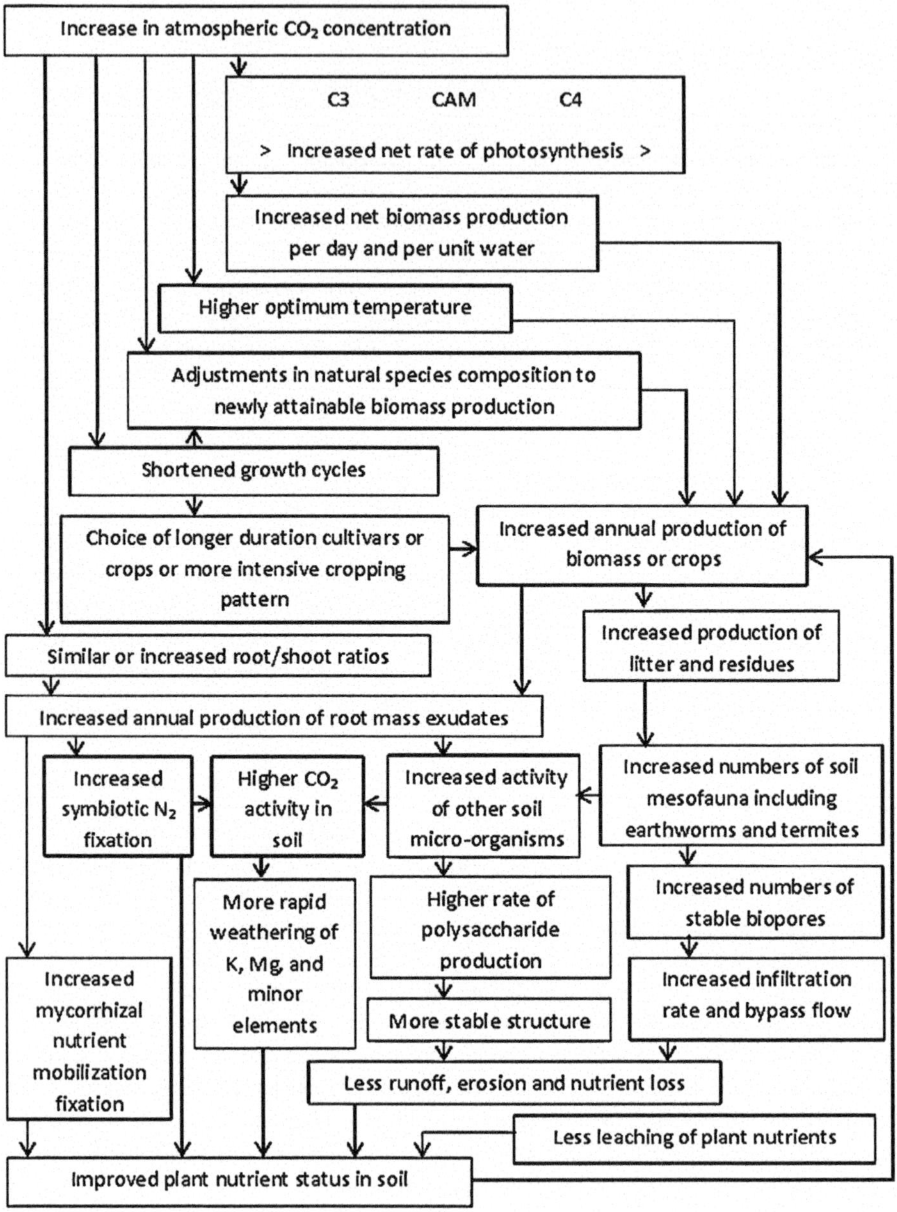

Fig. 3.2 Effect of increased atmospheric CO_2 concentration in soil quality (cited from Brinkman and Sombroek 1996)

agricultural production. Change in climate is also expected to change (increase) the types, frequencies and intensities of various crop and livestock pests, which in most of the cases are overlooked (Fischer et al. 1996; Reilly 1996; Adams et al. 1998; Easterling et al. 2007). Therefore, the impacts of climate on agriculture through weeds, insects, and diseases are most often understood qualitatively and quantitative knowledge is lacking (Tubiello et al. 2008). Changes in agronomic pests would be one of the first noticeable effects of climate change as it will potentially affect the pest-host relationship in following three ways (Tinker et al. 1996):

1. By affecting the pest population,
2. By affecting the host population, and
3. By affecting the pest–host interaction.

Therefore, its net effect might be noticed in several ways, which could be as follows (Tinker et al. 1996):

1. Pests that are of minor significance at present may become key in future when climate changes thereby causing serious losses,
2. The distribution and intensity of current key pests may be affected, leading to changed effects on yield and also on mitigation techniques such as pesticides and integrated pest management. For instance, inter kill of pests is likely to be reduced at high latitudes, resulting in greater crop losses and higher need for pest control (Sombroek and Gommes 1996), and
3. The competitive abilities in weed–plant interactions may be affected through changes in eco-physiology.

Usually, with the rise in atmospheric CO_2 average temperature will also increase and precipitation will become more variable. Plant's responses to such environmental stresses will determine pest infestation because pests often respond to plant responses such as changed plant physical quality, plant communities, and vegetation structures. Details of the whole insect physiological responses to changes in plant physiology as well as climate change are presented by Joern et al. (2005). Therefore, changed temperatures and thereby changed food quality will likely be primary drivers affecting agricultural pests. In this way, the climate change effect on pests may be initially understood as responses by plants that are then tracked by pests (Joern et al. 2005). For instance, changing environmental conditions will alter the competitive advantage of one species over another in a given system, thus there is evidence that there is increasing competition between C_3 crops and C_4 weeds thereby reduction in yield of the crop (Ziska and George 2004). Besides, there are also chances that the important interspecific interactions that check insect herbivores by a number of natural enemies would become vulnerable due to climate change. In addition, increase in climate extremes affect the physiological state of the host plant, and significantly alter patterns of resource allocation among plant tissues making it more susceptible to pest outbreak (Gan 2004).

Such increased pest infestations adversely affect plant and animal production through critical damage on plant tissues and transfer of diseases that affect the quantity as well as quality of food production (Joern et al. 2005). It has been established that Banks Grass Mite populations are favored by warm dry condition, whereas the predatory mite is favored by cool humid condition. Therefore, a short period, even a day, of hot and dry weather is sufficient for an outbreak of Banks Grass Mite in corn (Joern et al. 2005) thereby affecting the yield as well as the quality of the corn. Similarly, Crozier and Dwyer (2006) found that warming trends has led to early insect activities in the spring and proliferation of some species like the mountain pine beetle in the USA and Canada. Studies have shown that animal pests are spreading at a significant rate from low to mid-latitudes as a result of the warming trend. The models have projected that Bluetongue disease mostly in sheep and occasionally in goat and deer will spread from the tropics to mid-latitudes (Anon 2006; Wuijckhuise et al. 2006). Such spread is already experienced with the first ever incidence detected in Northern Europe in 2006, followed by a major outbreak in the subsequent years (Jean-Francois et al. 2007; Meiswinkel et al. 2008).

Similarly, White et al. (2003) simulated that climate change will increase the vulnerability of the Australian beef industry to the cattle tick (*Boophilus microplus*). Climate change in the form of extreme weather events can also result in catastrophic losses in confined cattle feedlots if prior conditioning is lacking. For instance, droughts in Africa during 1981–1999 have induced mortality rates of 18–92 % of livestock species (Easterling et al. 2007). However, almost all the studies on the impact of climate change on pest infestation is focused on the case of developed countries, whereas in the case of developing countries, which is also exposed to a higher degree of vulnerability to climate change such studies are virtually missing.

3.4 Animal Physiology

It is not only that animal production will be affected through the dynamics in soil carbon that determines the pasture quality, and insect infestation, but also the change in climate will have direct bearing on animal physiology that hampers livestock production. It has been established that the onset of a thermal challenge often results in decline in physical activity with associated declines in feed intake (through stall feeding as well as grazing) (Mader and Davis 2004). This consequently puts a ceiling on dairy milk yield. This ceiling ranges between 33 % and 50 % of the potential of Friesians cow breeds (Parsons et al. 2001). In addition, the energy deficit will be more prominent with the start of lactation. Consequently, there will be decreased cow fertility, fitness, and longevity (King et al. 2005). Conception rates of domestic animals are other important aspects of animal physiology going to be affected by increases in air temperature and/or humidity. Such problem is more prevalent among the cattle for which the primary breeding season occurs in the spring and summer months. Also, the hatchability of poultry is expected to be adversely affected by a rise in temperature (Easterling et al. 2007).

References

Adams RM, Hurd BH, Lenhart S, Leary N (1998) Effects of global climate change on agriculture: an interpretative review. Clim Res 11:19–30

Allard V, Newton PC, Lieffering M, Soussana JF, Grieu P, Matthew C (2004) Elevated CO_2 effects on decomposition processes in a grazed grassland. Glob Chang Biol 10(9):1553–1564

Anon. (2006) Exotic diseases: Bluetongue confirmed in France. Vet Rec 159:331

Bongaarts J (1994) Can the growing human population feed itself? Sci Am 270:36–42

Brinkman R, Sombroek WG (1996) The effects of global change on soil condition in relation to plant growth and food production. In: Bazzaz F, Sombroek WG (eds) Global climate change and agricultural production: direct and indirect effects of changing hydrological, pedological and plant physiological processes. Wiley, West Sussex, pp 49–63

Ciais P, Reichstein M, Viovy N, Granier A, Ogee J, Allard V, Valentini R (2005) Europe-wide reduction in primary productivity caused by the heat and drought in 2003. Nature 437:529–533

Crozier L, Dwyer G (2006) Combining population-dynamic and ecophysiological models to predict climate-induced insect range shifts. Am Nat 167:853–866

Easterling WE, Aggarwal PK, Batima P, Brander KM, Erda L, Howden SM, Tubiello FN (2007) Food, fibre and forest products. In: Parry ML, Canziani OF, Palutikof PJJP, Hanson CE (eds) Climate change 2007: impacts, adaptation and vulnerability. Contribution of Working Group II to the Fourth Assessment Report of the Intergovernmental Panel on Climate Change. Cambridge University Press, Cambridge, pp 273–313

Fischer G, Frohberg K, Parry ML, Rosenzweig C (1996) The potential effects of climate change on world food production and security. In: Bazzaz F, Sombroek W (eds) Global climate change and agricultural production: direct and indirect effects of changing hydrological, pedological and plant physiological processes. Wiley, West Sussex

Furbank RT, Taylor WC (1995) Regulation of photosynthesis in C3 and C4 plants: a molecular approach. Plant Cell 7:797–807

Gan JB (2004) Risk and damage of southern pine beetle outbreaks under global climate change. For Ecol Manage 191(1–3):61–71

Gill RA, Polley HW, Johnson HB, Anderson LJ, Maherali H, Jackson RB (2002) Nonlinear grassland responses to past and future atmospheric CO_2. Nature 417(6886):279–282

Jean-Francois T, Sailleau C, Mast J, Houdart P, Czaplicki G, Demeestere L, Clercq KD (2007) Bluetongue in Belgium, 2006. Emerg Infect Dis 13(4):614–616

Joern A, Logan JD, Wolesensky W (2005) Effects of global climate change on agricultural pests: possible impacts and dynamics at population, species interaction, and community levels. In: Lal R, Stewart BA, Uphoff N, Hansen DO (eds) Climate change and global food security. Taylor & Francis Group, Boca Raton, pp 321–362

King JM, Parsons DJ, Turnpenny JR, Nyangaga J, Bakari P, Wathes CM (2005) Ceiling to milk yield on Kenya smallholdings requires rethink of dairy development policy. In: Proceedings of the British Society of Animal Science. British Society of Animal Science, United Kingdom, p 25

Lal R (2005) Climate change, soil carbon dynamics, and global food security. In: Lal R, Stewart BA, Uphoff N, Hansen DO (eds) Climate change and global food security. Taylor & Francis Group, Boca Raton, pp 113–142

Loya WM, Pregitzer KS, Karberg NJ, King JS, Giardina JP (2003) Reduction of soil carbon formation by tropospheric ozone under increased carbon dioxide levels. Nature 425:705–707

Mader TL, Davis MS (2004) Effect of management strategies on reducing heat stress of feedlot cattle: feed and water intake. J Anim Sci 82:3077–3087

Meiswinkel R, Baldet T, Deken R, Takken W, Delécolle J-C, Mellor PS (2008) The 2006 outbreak of bluetongue in northern Europe—the entomological perspective. Prev Vet Med 87(1–2):55–63

Mendelsohn R, Dinar A (2009) Climate change and agriculture: an economic analysis of global impacts, adaptation, and distributional effects. Edward Elgar, Cheltenham

Morison JI (1996) Climate change and crop growth. Environ Manage Health 7(2):24–27

Nearing MA, Pruski FF, O'Neal MR (2004) Expected climate change impacts on soil erosion rates: a review. J Soil Water Conserv 59:43–50

Nelson M, Mareida M (2001) Environmental impacts of the CGIAR: an assessment. Consultative Group for International Agricultural Research, Durban

Oldeman L, Hakkeling RT, Sombroek WG (1991) World map of the status of human-induced soil degradation: an explanatory note. The Netherlands: Global Assessment of Soil Degradation, Wageningen. Retrieved 17 November 2011, from http://www.isric.nl/ISRIC/webdocs/docs/explannote.pdf

Parsons DJ, Armstrong AC, Turnpenny JR, Matthews AM, Cooper K, Clark JA (2001) Integrated models of livestock systems for climate change studies. 1. Grazing systems. Glob Chang Biol 7:93–112

Reardon T (1995) Sustainability issues for agricultural research strategies in the semi-arid tropics: focus on the Sahel. Agr Syst 48(3):345–359

Reilly J (1996) Climate change, global agriculture and regional vulnerability. In: Bazzaz F, Sombroek W (eds) Global climate change and agricultural production: direct and indirect effects of changing hydrological, pedological and plant physiological processes. Wiley, West Sussex

Rhodes ER (1995) Nutrient depletion by food crops in Ghana and soil organic nitrogen management. Agr Syst 48(1):101–118

Rosenzweig C, Hillel D (2005) Climate change, agriculture, and sustainability. In: Lal R, Uphoff N, Stewart BA, Hansen DO (eds) Climate change and global food security. Taylor & Francis Group, Boca Raton, pp 243–268

Scherr S, Yadav S (1996) Land degradation in the developing world: implications for food, agriculture and the environment to 2020. Food, Agriculture and the Environment Discussion Paper 14. International Food Policy Research Institute, Washington, DC

Sombroek WB, Gommes R (1996) The climate change–agriculture conundrum. In: Bazzaz F, Sombroek W (eds) Global climate change and agricultural production: direct and indirect effects of changing hydrological, pedological and plant physiological processes. Wiley, West Sussex, pp 1–14

Stoorvogel JJ, Smaling EM (1990) Assessment of soil nutrient depletion in Sub-Saharan Africa. The Winand Staring Center for Integrated Soil and Water Research, Wageningen

Tinker B, Goudriaan J, Teng P, Swift M, Linder S, Ingram J, Geijn SV (1996) Global change impacts on agriculture, forestry and soils: the programme of the global change and terrestrial ecosystems. In: Bazzaz F, Sombroek W (eds) Global climate change and agricultural production: direct and indirect effects of changing hydrological, pedological and plant physiological processes. Wiley, West Sussex, pp 295–318

Tubiello F, Schmidhuber J, Howden M, Neofotis PG, Park S, Fernandes E, Thapa D (2008) Climate change response strategies for agriculture: challenges and opportunities for the 21st century. Agriculture and Rural Development Discussion Paper. The World Bank, Washington, DC

UNEP & UNFCCC (2002) Climate change information sheet 10: agriculture and food security. In: Climate change information kit. United Nations Environment Programme (UNEP) & United Nations Framework Convention on Climate Change (UNFCCC), Switzerland

UNEP (1997) Global environment outlook. United Nations Environment Programme (UNEP) and Oxford University Press, Oxford

van Ittersum MK, Howden SM, Asseng A (2003) Sensitivity of productivity and deep drainage of wheat cropping systems in Mediterranean environment to changes in CO_2, temperature and precipitation. Agric Ecosyst Environ 97(1–3):255–273

White N, Sutherst RW, Hall N, Whish-Wilson P (2003) The vulnerability of the Australian beef industry to impacts of the cattle tick (*Boophilus microplus*) under climate change. Clim Chang 61:157–190

Wiliams A, Pruski FF, Nearing MA (2002) Indirect impacts of climate change that affect agricultural production: Soil erosion. In: OC III, Randolph JC, Southworth J, Pfeiffer RA (eds) Effects of climate change and variability on agricultural production systems. Kluwer Academic, Dordrecht, pp 249–264

Wuijckhuise V, Dercksen LD, Muskens J, Bruyn J, Scheepers M, Vrouenreats R (2006) Bluetongue in the Netherlands; description of the first clinical cases and differential diagnosis; common symptoms just a little different and in too many herds. Tijdschr Diergeneesk 131:649–654

Ziska LH, George K (2004) Rising carbon dioxide and invasive noxious plants: potential threats and consequences. World Resource Rev 16:427–447

Chapter 4
Agriculture in International Climate Change Negotiations

Abstract This chapter deals with the importance of agriculture in the economy of developing countries and discusses the link between agriculture and climate change covering the issue of adaptation, mitigation and resilience in agriculture. Similarly, this chapter also discusses how these aspects can be related to international climate change negotiations and also explores the potential benefits for poor farmers in developing countries through these negotiations. Likewise, the chapter also deals with the cross cutting issues of in agriculture and climate change focusing on cross cutting issues in adaptation and mitigation, governance and gender.

Keywords Adaptation • Governance • Mitigation • Resilience • UNFCCC

4.1 Introduction

4.1.1 Background

There is no doubt that human has impacted in making the rate of climate change faster. The Intergovernmental Panel on Climate Change (IPCC) has predicted that there will be an increase in temperature, rise in sea level, increase in frequency of extreme weathers due to climate change which will impact on food production, health, biodiversity, land etc. (FAO 2008a). The consequences of climate change will be difficult to predict but those societies that are already suffering severe development stress will be the most heavily affected and in particular, the most vulnerable sections of those societies (Jordan 2009). So, to minimize global instabilities caused by climate change induced crises, it is essential that the international community invest in mitigation, adaptation and resilience of the community especially those identified as being the most vulnerable (Jordan 2009). The section of the communities in these developing countries that are the most vulnerable to climate

K.L. Maharjan and N.P. Joshi, *Climate Change, Agriculture and Rural Livelihoods in Developing Countries*, Advances in Asian Human-Environmental Research,
DOI 10.1007/978-4-431-54343-5_4,

change are those, which are directly dependent on climatic factors like agriculture which is the main livelihood option for most of the rural population in developing countries (Department: Science and Technology 2007). As this section will be impacted highly by the changing climate, it will also require greater adaptation effort and capacity for building resilience (ICTSD and IPC 2009). At the same time agricultural sector is also responsible for a greenhouse gas emission, so agriculture has an important role in climate change mitigation also. This important role of agriculture in climate change should be seen from the local and global level. Though the importance of agriculture in the world and in relation to climate change has been understood and recognized, it has not figured largely in the international climate change negotiations to date. However, there is hope that it will figure more prominently in the future (ICTSD and IPC 2009). According to the Meridian Institute (2011) some of the key features of agriculture are as follows:

- Agriculture is a food producing sector which contributes to the basic needs of people and is one of the major sources of livelihood in developing countries.
- Agriculture is highly site and context specific, so technologies in one area may be synergistic while it may have a detrimental effect in another due to climate change.
- Agriculture is affected by climate change and climate is reversely affected by agriculture but the sector has great potential for synergies among mitigation, adaptation, food security and poverty reduction.
- As climate change threatens food production and supply, adaptation measures and resilience becomes very expensive in the future and also there might be trade-offs between food security and mitigation.
- The mitigation potential of agriculture is in the sequestration of carbon even while providing food security. The worldwide agricultural production offers considerable mitigation possibilities that—with an estimated potential of 5.5–6 Gt CO_2-eq per year—is almost equal to its current total annual emissions (5.1–6.1 Gt CO_2-eq).
- Agriculture will bring complex links between the issues of climate change and food security as climate change will likely affect agricultural production, distribution and supply of food and alter food prices.

Also, at the same time agriculture is one of the principle drivers of deforestation so it creates new linkages between it and emission reduction from deforestation.

4.1.2 Link Between Agriculture and Climate Change

The population of the world is increasing at a rapid rate, which will increase demands on land for food and fuels. For this there is need of increase in agricultural production but with climate change there will be added pressure on agriculture. Agriculture will be affected by both long-term mean temperature, precipitation and wind trend as well as by climate variability according to the Fourth Assessment Report of the Intergovernmental Panel on Climate Change (Meridian Institute

2011). In addition, agriculture will also be affected by the elevated carbon dioxide concentration, increase in weeds, pest and disease pressure (FAO 2009). The climate change will have mostly a detrimental effect on agriculture but different regions will have different degree of effect. The agriculture sector in Sub-Saharan region is to be negatively impacted most at 56 % followed by Asia at 26 %. Further, livestock production will also be affected directly as well as indirectly, directly such as physiological stress due to temperature increase and indirectly through changes in fodder quality and availability (Meridian Institute 2011). In addition to this, climate change will have an effect on the water resource availability, which is the major input for agriculture. As water resources will become scarce it will impact on the irrigation potential of the area and finally affecting the agriculture.

Agriculture is mostly negatively impacted by climate change but at the same time it is also one of the contributors of the greenhouse gases. According to the IPCC Fourth Assessment Report agriculture sector accounts 10–12 % of the total anthropogenic GHG emissions. The major agricultural emission includes methane (CH_4) and nitrous oxide N_2O which accounts 47 % and 58 %, respectively of the global emission. The agricultural emission constitutes N_2O emission from soil 38 %, CH_4 from enteric fermentation 32 % and the rest constitutes carbon dioxide from biomass burning 12 %, rice production 11 %, and manure management 7 % (Smith et al. 2007a). So, in agricultural emission, carbon dioxide only accounts for a small portion which is emitted as well as absorbed by agricultural soil resulting net emission of 40 megatons (Mt) CO_2-eq, less than 1 % of global anthropogenic CO_2 emissions (Meridian Institute 2011). As agriculture sector is one of the contributors of the greenhouse gases, it also has mitigation opportunities for reduction of greenhouse gases by reducing its emission as well as acting as a sink for carbon sequestration.

4.1.3 International Climate Change Negotiations and Links to Agriculture

The issue of climate change has been discussed in the international community for the past few decades. In 1992, countries came together for the international treaty on climate change called the United Nations Framework Convention on Climate Change (UNFCCC) so that they could limit the global temperature increase as well as cope with the impacts. Then in 1997 countries adopted the Kyoto Protocol, which set binding commitments for developed countries to reduce their emissions. The UNFCCC supports all the institutions and bodies involved in the international climate change negotiations (UNFCCC 2011). UNFCCC Article 2 focuses on the food production and sustainable development by stabilizing greenhouse gases. It states that GHGs level will be achieved to such a level within a time frame so that ecosystems can be adapted naturally to climate change ensuring food production is not threatened and enabling sustainable economic development. Similarly, UNFCCC Article 4.1.c states that there should be cooperation and promotion in development application through technology transfer, processes and practices for reductions of greenhouse gases, which also includes the emissions from agriculture and forestry.

UNFCCC Article 4.1.e refers to the cooperation in adaptation for the impact of climate change especially for water resources and agriculture (UN 1992).

Agriculture is mentioned in the Kyoto Protocol in Article 2.1.a.III, which states that the sustainable form of agriculture should be promoted in light of climate change. Also, the Kyoto Protocol binds the industrialized country for a reduction in GHGs by 2012 and also gives mainly three mechanisms for meeting their target for GHG emissions reduction, namely Emission Trading, Clean Development Mechanism, and Joint Implementation. Further, in Article 10.b.i of the Kyoto Protocol, it is stressed that the government should publish, update and implement programs for adaptation and mitigation at national and where possible regional level in the sector of agriculture and forestry (UN 1998). While the Clean Development Mechanism (CDM) maintains focus only on emission reductions rather than removals, the mitigation opportunities for small farmers in developing countries through cropland or rangeland management are excluded. There is no section dedicated to agriculture in Cancun agreements, sixth session of the Conference of Parties (COP) of the UNFCCC and sixth session of COP and Meeting of Parties (MOP) for Kyoto Protocol (COP/MOP) (Meridian Institute 2011).

In Cancun agreement, though there is no decision or work program dedicated to agriculture but has been taken as driver of deforestation, so agriculture can be considered under mitigation processes. Also, according to the Cancun agreement, the least-developed countries are to provide National Adaptation Programmes of Action (NAPA), which prominently figured agriculture. Developing countries indicated to undertake nationally appropriate mitigation actions related to agriculture like restoration of grasslands, fodder crop productions etc. following fifteenth session of COP. Besides involving in international negotiations on different issues of adaptation, mitigation and reduction of GHG like financing and support mechanisms and methodologies, countries can take a lead in developing capacities, confidence in engaging themselves with agricultural adaptation and mitigation along with food security and development. Also, the government could initiate making policies that will help agriculture to cope with the climate change, food security and their developmental goal by integrating agriculture with land-use planning strategies (Meridian Institute 2011). Besides these the few important points in climate change regime and agriculture as per International Centre for Trade and Sustainable Development & International Food and Agricultural Trade Policy Council (2009) are as follows:

- All the countries are necessary to submit national communication plans which should include emission from all sectors including agriculture and policies to mitigate it.
- The industrialized countries that are Annex I countries are required to provide annual quantitative inventories regarding their emission including agriculture.
- The Guidelines for National GHG inventories as given by IPCC 2006 proposed to integrate agriculture, and Land Use and Land Use Change and Forestry (LULUCF) guidance reporting into one sector the Agriculture, Forestry and Other Land Use (AFOLU).
- Providing mitigation for the AFOLU sector for non-Annex I countries have not been tapped under the existing regime.

4.2 Adaptation, Mitigation and Resilience in Agriculture

4.2.1 Adaptation

Adaptation is an important issue in order to reduce the adverse impact of climate change and to increase the resilience to it (UNFCCC 2011). According to the UNFCCC (2011) the issues of adaptation that are being highlighted in different parties under different convention bodies are as follows:

- The Cancun adaptation framework objective was to improve action on adaptation through international cooperation and giving proper attention to adaptation matter related to the convention.
- The Nairobi work program's objective was to help countries improve understanding of climate change impact and to make informed decision on practical adaptation actions.
- Implementation of the programs related to adaptation like NAPA, and helping under subsidiary body for implementation through finance, technology and capacity building.

There are no any globally applicable adaptation measures as it depends from region to region since the impact of climate change will vary according to region. Therefore, it is crucial to do country-specific research for any adaptation measures on agricultural practices (Meridian Institute 2011).

According to FAO (2008b), in agriculture sector there is high potential of synergies between climate change adaptation and sustainable development. The adaptation measures to climate change impact will not just help to adjust to climate change but also will help in socio-economic development. So, adaptation is an important issue for agricultural sector policies of developing countries as well as for overseas development assistance and thus adaptation should be given more emphasis in new international climate change regime (ICTSD and IPC 2009).

Adaptation and Food Production

Food production is one of the major challenges that we will face in the coming future with climate change affecting the agriculture. The adaptation practices and effective adaptation policies will help to increase farming capacity and food system in the face of climate change and also maintain food production. With climate change there will be changes in both mean temperature and precipitation that will cause variability between seasons and also increases the frequency of natural disasters and extreme events like droughts, floods. This shift in weather patterns will hamper the agriculture. Still there are some uncertainties about the degree of impact in many of the world's agriculture (Meridian Institute 2011). Further, many projected impacts of climate change studies on agriculture are the extension and intensification of the challenges of climate variability that are already there especially for rain-fed cropping and livestock system (Vermeulen et al. 2010). As there are still uncertainties regarding the degree of impacts in different areas, the adaptation intervention should be focused according to the area under consideration. Hence, for adaptation in agriculture especially in rural area of developing countries there is need of no-regret strategies regardless of direction or magnitude of change (ICIMOD 2010). Also, the most effective adaptation interventions are generally built on current practices and technologies as farmers have generations of experience in managing climatic risks (Meridian Institute 2011).

According to the IPCC, adaptation has been distinguished as autonomous adaptation, that is, response, which is triggered by ecological changes in natural systems and by market welfare changes in human systems and planned adaptation which is the result of the deliberate policy change (IPCC 2001). These both farm-level and higher-level policies and investments adaptation will be necessary for effective agricultural adaptation. Along with the adaptation in farming practices, there are significant scope for agricultural adaptation throughout the food chain, such as better post-harvest storage and distribution of food, in order to amend the gap between good and poor harvest years (Meridian Institute 2011).

Adaptation can occur at various levels from the farm level to the policy level, like changes in agricultural practices, varietal change, substitution or diversification, moving out of crop farming, aquaculture, or livestock rearing. For the sustainability of adaptation it should be technologically viable, socially acceptable and economically feasible (Vermeulen et al. 2010). At autonomous adaptation level various options for cropping and livestock systems are described as follows (Smith et al. 2007a):

- Different varieties or species with greater resistance,
- New cropping practices, including timing and area for crop production, and improve water and fertilizer management,
- Greater use of water conservation technologies,
- Diversification of on-farm activities and enhancement of agro-biodiversity,
- Adapted livestock and pasture management, and
- Improved management of pests, diseases and weeds.

The adaptation measures are mainly area specific and also there are no globally applicable measures that can be prioritized, hence interventions should be locally selected and should be appropriate. Also as there are many limitations to many of the practices the cross-site and cross-country transfer of best practices will be critical to effective adaptation. Further, there is also need for the continuous advancement in the technology for adaptation to the climate change. Also, investment in the infrastructure sector and sustainable development of the rural community will be essential for any community to undertake the adaptation measures. The adaptation intervention should be adopted based on the holistic approach, i.e., intervention should support agriculture and food systems, which contributes to livelihoods, income and food security (Meridian Institute 2011).

4.2.2 *Mitigation*

Agriculture, one of the sectors that will be highly impacted by climate change, has huge potential for mitigation. But mitigation options are coupled with difficult challenges, which are mainly due to the huge size of land area under agriculture in the world (at the same time it is also the breadth of opportunity for huge carbon sequestration), different agro-ecosystems and farming systems, and involvement of a large number of farmers (FAO 2008b). Though mitigation is difficult and challenging, agriculture does have huge technical mitigation potential, which has been estimated at 5.5–6 Gt of CO_2 per year by 2030, if best management practice is widely adopted (Smith et al. 2007a). The reduction in CO_2 is the highest from soil carbon sequestration (89 %), followed by emissions reductions of methane (9 %) and N_2O (2 %) (FAO 2008b). Further, agriculture provides an important and relatively cost-effective mitigation options, hence it should be addressed fully by climate change regimes especially in the agricultural measures of developing countries (ICTSD and IPC 2009). Hence, developing countries mitigation through agriculture could be a feasible way to contribute to climate change regime, however minimal progress has been made in this sector for developing countries (Murphy et al. 2010). In general, agriculture and small farm holders have huge potential to mitigate GHG emissions and generating co-benefits for sustainable development.

Mitigation and Food Production

Agriculture is one of the sectors that contribute to the global GHGs and it also is a potential sector for reducing GHG emission drastically. In the climate change regime mitigation in agriculture has mainly been focused. The biggest challenge of mitigation in agriculture is achieving it without compromising food security both nationally and globally (Meridian Institute 2011). Developing countries are to play an important role in GHGs emissions mitigation through agriculture due to growing agriculture and forestry emission, and also provide the largest and most cost-effective mitigation opportunities (Murphy et al. 2010). Though developing countries have the mitigation potential it will be their biggest challenge, because with the increase in agriculture there will be an increase in GHGs from the sector as well. However, in agriculture there is still mitigation potential for GHGs emission reduction, which can be done mainly by increasing efficiency in agriculture production and also through reduction in emission along with removal of carbon through sequestration in agricultural soils and biomass (Murphy et al. 2010). Also, there is co-benefit of emission reduction and increasing food productivity with the advancement of the technology. According to the Fourth assessment report of the IPCC, the mitigation options from agriculture can be broadly distinguished into seven categories (Smith et al. 2007a) which are as follows:

- Cropland management,
- Grazing land management and pasture improvement,

- Management of organic soils,
- Restoration of degraded lands,
- Livestock management,
- Manure management, and
- Bioenergy.

These mitigation options have a high potential for synergies with food security like improved crop varieties, not keeping the land fallow, crop rotations, fertilizer management, improving forage quality and quantity on pastures, increasing energy-efficient irrigation and water conservation techniques, and implementing agroforestry (Meridian Institute 2011). These mitigation options will have co-benefit of increased food production as well as mitigate GHGs emissions especially in the developing and least developed countries as majority of people in these countries depends on agriculture for their livelihood. However, there are still barriers in the current climate change regime to implement mitigation options, some of the barriers as given by Smith et al. (2007b) are:

- Permanence of carbon sequestration,
- Identifying the additional mitigation option to ongoing activities in the absence of a market, and
- Uncertainty of mechanism and measurement like uncertainty of complex biological and ecological processes of trace gas emission and carbon storage in agricultural system.

Other than these, there are few other constraints like cost-effective and appropriate mitigation options at global level, i.e. until the farmers feel the mitigation options are benefiting them they will not adopt it, and mitigation options may not increase food productivity in all the regions (Smith et al. 2007b). Hence, mitigation in agriculture should be locally appropriate and in some cases technically feasible innovation may be required. For mitigation intervention in any region, along with innovation in technology things like food security, synergies and tradeoffs of mitigation intervention in the region, and climate financing should also be considered. Further, some mitigation options may have negative impacts on short-term but in the long-term may prove to have positive impacts for increasing both the average and stability of production levels in agriculture which also needs to be focused while considering mitigation intervention (Meridian Institute 2011).

4.2.3 Resilience

Resilience is defined in various ways by different literatures. According to the review done by Jordan (2009) different literatures describe resilience as follows:

- Ability of a community to anticipate and prepare for risk and adapt before the event occurs.
- Ability of a system to recover after the damage has been done.
- Ability of a system to cope with the hazards and also to make adjustments in response to a hazard.

Jordan (2009) also described the resilience from layered concept from individual to global scale and added that resilience at one level may not necessarily create resilience at another level. In addition to this, Resilience Alliance (2012) stated that resilience has three distinct characteristics, namely:

- The amount of change the system can undergo and still retain the same controls on function and structure,
- The degree to which the system is capable of self-organization, and
- The ability to build and increase the capacity of learning and adaptation.

According to the UNFCCC, technical support to adaptation is one of the key priorities to improve the resilience of agriculture against adverse impacts of climate change. Also, increasing resilience in the agriculture sector from the impact of climate change will also alleviate poverty and increase food production. Therefore, policies should be targeted at improving agricultural productivity, rural physical and institutional infrastructure, and poverty alleviation in rural areas (ICTSD and IPC 2009).

Resilient Farming Needs

Farmers have lots of experience in dealing with the climatic condition and agriculture, therefore their experience and knowledge should be shared across the region and countries for resilient farming. According to the Kenny (2010) the experiences of farmers for making resilient farming are as follows:

1. There is a need for raising awareness and information regarding climate change and associated issues as there is a lack of understanding in relation to potential co-benefits of adaptation and mitigation from range of actions.
2. Further, as climate change is a multi-disciplinary field, there is a need of coordination between all the concerned parties to work together.
3. Farmers' innovativeness and research should be supported strongly and lessons should be learned from it. Farmers who have shown innovativeness (e.g., organic, biological and low-input farmers) should be focussed. Also a closer understanding of the fundamentals of ecology and resilience in farming systems itself will help with adaptation and mitigation co-benefits
4. The sustainability and resilience of the farming should be considered. Though profitability remains important to farmers, the issues like long-term sustainability of some practices and development of greater resilience should also be focused.

The improvement in resilience may have co-benefits for the farmers as well as climate change regimes. Some of the co-benefits from increasing resilience as given by Asian Development Bank (ADB) & International Food Policy Research Institute (IFPRI) (2009) are as follows:

- Increased adaptation of crops and livestock to climate stress,
- Enhanced access and utilization of technology and information,
- Increased income generation,
- Increased use of resource conserving technologies,
- Open and transparent trade regimes, and
- Improved risk sharing

4.3 Cross Cutting Issues in Agriculture and Climate Change

4.3.1 Cross Cutting Issues in Adaptation and Mitigation

The core issues with the agricultural adaptation and mitigation is producing more food efficiently under a volatile climatic condition with a net reduction in GHGs emission from food production and marketing (Lybbert and Sumner 2010). Hence, there is a need of technological advancement and innovativeness for adaptation and mitigation options in agriculture. These are also feasible for farmers both from the viewpoint of economic as well as social perspectives. Further, to cope with the climate change there is a need for new dimensions of agricultural cross cuttings policies and technologies for mitigation and adaptation in agriculture. For this purpose, the government in their locality have to support their local innovative issues and long-term adaptation and mitigation plans and policies in the area. Also, it becomes very important that we should think about the sustainable agriculture, which just not contributes to production but also contributes to the livelihood of the farmers as well as the mitigation and adaptation issues of climate change.

Technologies of Agricultural Adaptation and Mitigation

There should be application of new innovative technologies, new land management practices, and water efficient technologies to deal with the long-term adaptation in order to reduce the adverse effect of climate change (FAO 2007). The adaptation involves implementation of a range of strategies for biotechnology, hard technology (machinery, equipment and tools), soft technologies (knowledge, capacity building, and awareness raising) and organizational technologies (institutional building), which can be from individual/household/community/institution level to national and international levels (Clements et al. 2011). Some of the adaptation technologies that can be implemented are as follows:

1. Changes in management practices: The changes in the management will include changes in the land use for maximizing yield, i.e., application of new technologies such as change in input use including organic and low external input agriculture, change in input like new varieties of crops and livestock, and change in planting dates (ADB and IFPRI 2009). Further, changes in production management and practices such as minimum tillage will be an effective way of adaptation (Lybbert and Sumner 2010).
2. Changes in agricultural water management: In the current scenario of increasing water demand along with climate change, improvement of water for agriculture and irrigation access should be efficiently managed (Lybbert and Sumner 2010). Water management techniques such as water harvesting, soil moisture conservation techniques, preventing water logging, erosion, and nutrient leaching are useful practical technologies. Further, water management includes the flexibility of irrigation system, which will have multiple uses like water for irrigation, urban

areas and industry, and the establishment of a better water control system at key distribution points (ADB and IFPRI 2009).

3. Agricultural diversification: One of the options for adaptation to the climate change is diversification of the farm level for example alternative rice/shrimp farming, also off farm employment and further organic farming and indigenous knowledge also calls for agricultural diversification (ADB and IFPRI 2009).

Besides these agricultural practices, another adaptation measures as pointed out by ADB and IFPRI (2009) are:

1. Development in science and technology of agriculture,
2. Agricultural information system for effective dissemination of modern technology to the farmers, and
3. Risk management and crop insurance.

In addition to adaptation, mitigation is also an important aspect for dealing with climate change and also with increasing magnitude of climate change adaptation will be costlier. So mitigation options should be followed beforehand when it is affordable (ADB and IFPRI 2009). Basically the mitigation falls in three broad characteristics, namely:

1. Reducing emission: The emission from agriculture can be reduced by managing efficiently the flows of carbon and nitrogen in agriculture which depend on local condition and vary from region to region (ADB and IFPRI 2009).
2. Enhancing removal: The proper management of the agricultural system and practices like Agroforestry can enhance the removal of atmospheric CO_2 in the form of soil organic matter (Lal 2004).
3. Avoiding emissions: The reduction of GHG emissions by using the crop residue as a source of fuel (Cannell 2003; Schneider and McCarl 2003) will put less pressure on the use of fossil fuel. Better crop management that helps in deforestation will also avoid the carbon emissions (Smith et al. 2008).

Further, emission of GHGs from agriculture comes basically from four sectors: agricultural soil, livestock and manure management, rice cultivation, and the burning of agricultural residue and clearing of forest and grassland (ADB and IFPRI 2009). For reduction of GHGs from these sources, the mitigation options as reviewed by the Smith et al. (2008) are as follows:

(a) Cropland management: The IPCC soil tool calculated carbon sequestration by cropland management of upper 30 cm of soil is about 50 % of the carbon stock in cropland soil. In the next 20 years, 0.5 Pg CO_2 can be sequestrated by changing cropland management (Schulze et al. 2006). This shows that cropland management can play a major role in carbon sequestrations, and it offers many opportunities for using practices to reduce GHGs emissions and some of them are suggested by Smith et al. (2007b) are as follows:

- Agronomy,
- Nutrient management,
- Tillage/residue management,

- Water management,
- Rice management,
- Agroforestry, and
- Land covers (use) change.

(b) Grazing land management and pasture improvement: Grazing land poses a great potential of carbon sequestration if it is managed properly (Smith et al. 2008). Some of the techniques for grassland management are as follows:

- Grazing intensity,
- Increased productivity (including fertilization),
- Nutrient management,
- Fire management, and
- Species introduction.

(c) Management of organic soils: Carbon accumulated over time remains in organic soil as a high density carbon which emits CO_2 and N_2O. These emissions can be reduced by practices like avoiding row crops and tubers, avoiding deep ploughing and maintaining shallow water table (Smith et al. 2007a).

(d) Restoration of degraded lands: Exacerbating the desertification by anthropogenic activities will reduce the soil productivity, jeopardize the food security, and impairs environment quality. Degradation of soil aggravates the GHGs emission but if desertification is controlled and degraded soil is restored it has potential of sequestrating about 0.9–1.9 Pg of carbon per year (Lal 2009).

(e) Livestock management: Methane from livestock sector, predominantly sheep and cattle, accounts for one-third of global anthropogenic gas of this gas (EPA 2006). The emission of GHGs can be reduced through following practices:

- Improved feeding practices,
- Specific agents and dietary additives, and
- Longer term management changes and animal breeding.

(f) Manure management: The significant amount of N_2O and CH_4 is released during storage, but the intensity varies according to the management practices (Smith et al. 2008). There are different ways of mitigating the GHGs from manure management. Some of the manure management techniques to mitigate GHGs are bio-digesters, gasification systems and composting (Langmead 2003).

(g) Bioenergy: The use of agriculture crop and residues are increasing as sources of feedstock for energy to replace fossil fuel. The bioenergy helps in mitigating GHGs as the CO_2 released by biofuel burning is of recent atmospheric origin (photosynthesis) and it displaces the CO_2 that would be emitted by fossil fuel (Smith et al. 2008). Still, there is large potential to exploit bioenergy for which major transition is needed (Smith et al. 2007a).

Synergies and Trade-Offs of Adaptation and Mitigation

Agriculture sector plays a dual role in climate change, in one hand as one of key sector that is affected by it and on the other hand as emitter of GHGs in the atmosphere (Rosenzweig and Tubiello 2007). In addition to this, agricultural system has to meet various objectives with the challenges from different aspects such as like climate change and policy has to balance potential synergies and tradeoffs of adaptation and mitigation measures in agriculture in order to meet food production and food systems. There are numerous opportunities for synergies in agriculture between adaptation and mitigation like conserving soil moisture, reducing soil degradation which can be obtained by adopting established and often low-tech good practices, and by taking up of agricultural innovations (Meridian Institute 2011). Also increased irrigation and fertilization in marginal semi-arid area may contribute in enhancing soil carbon in that area (Rosenzweig and Tubiello 2007). Further, the adaptation and mitigation have synergies with the food production especially in increasing crop yield by effective crop management (Branca et al. 2011). In addition to this, adaptation and mitigation intervention has additional benefits like greater biodiversity, and higher incomes to the farmers (Meridian Institute 2011). Some of the technologies that mitigate and also build resilience for farmers are adoption of high-yielding varieties, shifting to rice–wheat production systems and alternate dry–wet irrigation, which will help to withstand harsh climate as well as mitigate GHGs emission (ADB and IFPRI 2009).

Contrary to it, there are also tradeoffs associated with the adaptation and mitigation options in agriculture. For example, nitrous oxide emission might be increased while reducing rice paddy methane emission through periodic drying (Meridian Institute 2011). Also some adaptation and mitigation actions might not be conducive with each other as well, like increasing cultivation in marginal area towards pole might lead to substantial soil carbon loss from previously undisturbed land (Rosenzweig and Tubiello 2007). Adaptive capacity may be negatively affected by few mitigation measures like using crop residues for biofuels, which can deplete soil carbon (Meridian Institute 2011). The mitigation measures like reducing emission from deforestation might have a negative impact on the livelihood of the rural people and also might hamper food security and sustainable development (FAO 2008a). Especially the tradeoffs occur when production land is taken out for the mitigation options, for example restoring of wetland can take the wetlands out of production permanently (Meridian Institute 2011).

There is a great potential that the trade-offs risks can be reduced by the promotion of diverse and flexible livelihood and food production strategy, planned adaptation of food security and risk reduction, and flexible and adaptable institutions (FAO 2008a). The tradeoffs especially due to the landscape management, point the importance of coordination between different sectors like forestry and agricultural policies. In addition to this there are many options that make the whole food system more resilient, which will also make it more efficient and cut down the emissions from agriculture indirectly. For example, conservation tillage helps in soil water retention and also sequestrates carbon below ground. So, the policies and planning need to look upon not just food security and adaptation but also mitigation in national and local level especially in the least developed countries where people are more vulnerable. The justification behind this is that synergies and tradeoffs tend to be local to farming systems and agro-ecological zones (Meridian Institute 2011).

4.3.2 *Governance as Cross Cutting Issue in Climate Change*

There are number of governance challenges while intervening in the food and agriculture sector that are characteristic of agriculture and rural sectors (Birner et al. 2010). Further, climate change will add more problems to governance in different sectors as it present new challenges in the area. For example, with irregular precipitation patterns there will be governance challenges for water management (water rights, externalities) especially in water scarce and drought prone areas (Theesfeld et al. 2011). In addition to this, there will be challenges in climate change governance also, which may be local, national or international. According to Meadowcroft (2009), there may be a number of climate change issues that provide challenges for the existing governance mechanism which are:

- Societal reach: It will be very challenging to consciously steer societal behavior of consumption and production pattern to reduce emission dramatically.
- Scientific uncertainty: There remain uncertainties regarding the sensitivity of climatic systems, regional impacts of climate change, and consequences of changes in the ecosystem.
- Distributional and equity linkages: Climate change will have an impact on different region differently and its response from different region will also be different. Equity has always been a governance issue domestically as well as for the international community, and climate change will add to further regional disparities, fuel poverty etc.
- Long time frames: As the climate system evolves over the decades, climate change and greenhouse gas emissions issues have been a longstanding problem. Therefore, the political and official tenure of few years will fit poorly with such issues.
- Global implication: The cause of climate change is a global issue, while impact will be felt by different region differently so coordinating international community together will be a great challenge.

Governance in Adaptation and Mitigation

Governance of adaptation requires knowledge of the regional and local climatic effects and planning to deal with expected human activities (Meadowcroft 2009). As there will be continuous change in the climate, adaptation measures should also need to evolve. Adaptation is not something that needed to be started from the scratch but it is an incremental process that is generally based on long experience of previous adaptations (Lambrou and Piana 2006). For successful adaptation, critical elements are required like enhanced scientific knowledge of regional and local climate change, knowledge on ecosystems and societal impacts, information on socio-economic impacts, which is challenging issue for local as well as national government especially in least developed countries (Meadowcroft 2009). Some of the governance measures for adaptation listed by Meadowcroft to be taken into consideration (2009) include:

- Climate change impacts should be addressed in local and national level planning processes like land use planning,
- Reports on national and regional adaptation and anticipated adaptation cost,
- Adaptation forum with key stakeholders for impacts and responses, and
- Integrating adaptation strategies in planning of agriculture and natural resource management.

There is great potential for mitigation of GHGs in agriculture. Therefore, the following points are needed for the governance of mitigation; understanding of emissions sources, cost-effective abatement potentials, and policy approaches (Meadowcroft 2009). The main barriers for agricultural mitigation measures as given by Wollenberg and Christine (2011) are as follows:

- Lack of clear financial incentives,
- Concern about constraining economic development, food security and trade,
- Credibility and value of agricultural offset credits that are hindered by slow progress toward cap and trade markets,
- Lack of capacity for measuring and monitoring GHG in most developing countries and also technical information delivery and structure of accountability, and
- Lack of information for farmers about benefits and liabilities associated with carbon market contracts as well as technical options for mitigation.

There is array of policy instruments that are available for encouraging mitigation but difficulties are in political will to implement them (Meadowcroft 2009). Though there are challenges in the governance of mitigation there are also opportunities that might be helpful for mitigation governance. For example, there are technical options available that can synchronize with the economic development, adaptation and food security and also aggregate projects can reduce transaction cost and help with investments (Wollenberg and Christine 2011). So mitigation is possible in very limited resources with 'no regrets' policies with little or negative economic cost (Meadowcroft 2009).

4.3.3 Gender Dimension in Climate Change

Climate change will have an impact on poor and vulnerable people especially the farmers in agriculture where there is a gender differentiation impact (Fogelberg 2010). The key dimensions of gender inequality are men and women's differential access to social and physical goods or resources (Brody et al. 2008). In addition, as the majority of the poor constitute women they will be disproportionately impacted by the climate change. Especially in rural areas of developing countries women are involved as the primary user and caregiver of natural resources, which puts them at risk of climate change. Further, they are kept away from the processes and decision making on the use, and management of natural resources (Gaye et al. 2009). Furthermore, according to Lambrou and Piana (2006) the control of resources within societies is shaped by age, physical ability, citizenship status, social/ethnic and cultural group and gender. This will raise the question of sustainability of any development in the region. Hence, for addressing sustainable development followings are the key proposed by (Birner et al. 2010):

1. Gender equality is basic human right
2. Many disparities in development come from gender inequality
3. Gender inequality can have significant impact on efficiency and welfare

As women's involvement in agriculture and natural resource management is the most common especially in developing countries, responsibility of adaptation is also likely to fall on them, i.e., finding alternative ways to feed their family (CIDA 2002). But in adaptation research, there has been limited attention given to the difference between men and women within the risk population (Gaye et al. 2009).

The impact of climate change on women varies between region and culture. For adaptation planning a concrete situation and individual need should to be taken into consideration (Röhr 2007). According to Röhr (2007) some common subjects that are affected through region and cultures are:

1. Care work, poverty reduction, and income generating activities: Women are responsible for securing food, water and also caring for children and the elderly. With climate change impact there will be depletion of resources for which they have to spend more time for collecting plants and cultivating their crops.
2. Health Impacts: With the increasing temperature there will be increasing diseases, which will make women more vulnerable because of their physical vulnerabilities. Also climate change will put extra pressure on their health as they need to carry water from distant area.
3. Climate related natural disaster: Natural disaster like flood, drought, and hurricanes will impact men and women differently. For example, in 1991 cyclones and flood in Bangladesh claimed the death of five times more women than men. This is mainly due to social and cultural norms, which did not allow women to leave their homes without male relatives.

Due to some statutory and customary laws, women have less access to credit and agricultural extension services that gives them less incentive in making investments in land rehabilitation and soil quality. Despite these hindrances it is seen that women are developing effective coping strategies to climate change (Brody et al. 2008). In the research project done by ActionAid and Institute of Developmental Studies (IDS) in Nepal, India and Bangladesh, it was seen that women farmers were adapting to the changing climate in order to secure their livelihoods. Also this research showed that the women were clear about their needs, such as crop diversification and agricultural practices, in order to adapt to the flood. This showed that women were clear about the types of intervention needed in order for sustainable agriculture (Brody et al. 2008). Therefore, for climate adaptation it is necessary to approach from gender-aware perspective as it will not just improve women's adaptive capacity but also will improve the household that they are part of (Fogelberg 2010).

Degradation of environment will put more burden on women as they are primarily responsible for the water collection and household gardens, with drought condition they will suffer more, so the primary losers should be involved in the recovery strategies and efforts. With regards to climate change mitigation, the gender sensitivity aspects of technology and energy requirement should be considered from the international perspective all the way to the household level. If they are not systematically considered it might result in poor suit of the new technologies and policies in regards to gender (Lambrou and Piana 2006).

References

ADB & IFPRI (2009) Building climate resilience in the agriculture sector of Asia and the Pacific. Asian Development Bank (ADB), Philippines

Birner R, Quisumbing AR, Ahmed N (2010) Cross-cutting issues: governance and gender. Bangladesh Food Security Investment Forum, Dhaka

Branca G, Lipper L, Mann W, McCarthy N (2011) Adaptation and mitigation in agriculture: synergies and trade-offs. In: FAO-WB expert group meeting, Food and Agriculture Organization of United Nations, Rome

Brody A, Demetriades J, Esplen E (2008) Gender and climate change: mapping the linkages. A scoping study on knowledge and gaps. University of Sussex, Brighton. Retrieved 1 December 2011, from http://www.bridge.ids.ac.uk/reports/Climate_Change_DFID.pdf

Cannell MG (2003) Carbon sequestration and biomass energy offset: theoretical, potential and achievable capacities globally, in Europe and the U.K. Biomass Bioenergy 24:97–116

CIDA (2002) Gender equality and climate change: why consider gender equality when taking action on climate change? Canadian International Development Agency (CIDA), Quebec

Clements R, Haggar J, Quezada A, Torres J (2011) In: Zhu X (ed) Technologies for climate change adaptation–agriculture sector. UNEP Risø Centre, Roskilde

Department: Science and Technology (2007) South Africa's climate change technology needs assessment. Department: Science and Technology, South Africa

EPA (2006) Global anthropogenic non-CO_2 greenhouse gas emissions: 1990–2020. United States Environmental Protection Agency, Washington, DC

FAO (2007) Adaptation to climate change in agriculture, forestry and fisheries: perspective, framework and priorities. Food and Agriculture Organization of the United Nations, Rome

FAO (2008a) Climate change adaptation and mitigation in the food and agriculture sector. HLC/08/BAK/1. Food and Agriculture Organization of the United Nations, Rome

FAO (2008b) Enabling agriculture to contribute to climate change mitigation. United Nations Framework Convention on Climate Change, Bonn. Retrieved 12 December 2011, from http://unfccc.int/resource/docs/2008/smsn/igo/036.pdf

FAO (2009) How to feed the world in 2050. Food and Agriculture Organization of the United Nations, Rome

Fogelberg T (2010) Gender consideration in climate change and agriculture. Retrieved 12 December 2011, from The Broker: Connecting Worlds of Knowledge: http://www.thebrokeronline.eu/en/Blogs/It-s-Down-2-Earth-Conference-blog/Gender-considerations-in-climate-change-and-agriculture

Gaye I, Dioné J, Laye O, Tutu K, Akol C (2009) Gender and climate change: women matter. Economic Commission for Africa, Addis Ababa
ICIMOD (2010) Rural livelihoods and adaptation to climate change in the Himalayas. INFORMATION SHEET #5/10. International Centre for Integrated Mountain Development (ICIMOD), Kathmandu
ICTSD & IPC (2009) International climate change negotiations and agriculture. Policy Focus No. 1. International Centre for Trade and Sustainable Development (ICTSD)-International Food and Agricultural Trade Policy Council (IPC) Platform on Climate Change, Agriculture and Trade, Geneva. Retrieved from http://ictsd.org/downloads/2009/11/international-climate-change-negotiations-and-agriculture.pdf
IPCC (2001) Climate change 2001: impacts, adaptation and vulnerability. Cambridge University Press, Cambridge
Jordan J (2009) Rethinking community resilience to climate change: does a social capital lens help? In: Development Studies Association conference 2009 "Current Crises and New Opportunities", Development Studies Association, United Kingdom
Kenny G (2010) Adaptation in agriculture: lessons for resilience from eastern regions of New Zealand. In: Nottage R, Wratt D, Bornman J, Jones K (eds) Climate change adaptation in New Zealand: future scenarios and some sectoral perspectives. New Zealand Climate Change Centre, Wellington, pp 56–69
Lal R (2004) Soil carbon sequestration impacts on global climate change and food security. Science 304:1623–1627
Lal R (2009) Soil carbon sequestration through desertification control. In: Soils, society, and global change. UNU/SCS/European Commission, Selfoss
Lambrou Y, Piana G (2006) Gender: the missing component of the response to climate change. Food and Agriculture Organization of the United Nations, Rome. Retrieved 14 November 2011, from http://www.fao.org/sd/dim_pe1/docs/pe1_051001d1_en.pdf
Langmead C (2003) Manure management and greenhouse gas mitigation techniques: a comparative analysis. Discussion paper C3-013. Climate Change Central, Calgary
Lybbert T, Sumner D (2010) Agricultural technologies for climate change mitigation and adaptation in developing countries: policy options for innovation and technology diffusion. Issue Brief No. 6. International Centre for Trade and Sustainable Development & International Food and Agricultural Trade Policy Council, Geneva
Meadowcroft J (2009) Climate change governance. Policy Research Working Paper 4941. The World Bank, Washington, DC. Retrieved from http://www-wds.worldbank.org/servlet/WDSContentServer/WDSP/IB/2009/05/19/000158349_20090519144015/Rendered/PDF/WPS4941.pdf
Meridian Institute (2011) Addressing agriculture in climate change negotiations: a scoping report. Meridian Institute, Washington, DC. Retrieved 1 January 2012, from www.climate-agriculture.org
Murphy D, McCandless M, Drexhage J (2010) Expanding agriculture's role in the international climate change regime: capturing the opportunities. International Institute for Sustainable Development, Manitoba
Resilience Alliance (2012). Resilience. Retrieved 12 January 2012, from The Resilience Alliance: http://www.resalliance.org/index.php/resilience
Röhr U (2007) Gender, climate change and adaptation. Introduction to the gender dimensions. Focal Point Gender, Environment, Sustainability, Berlin
Rosenzweig C, Tubiello FN (2007) Adaptation and mitigation strategies in agriculture: an analysis of potential synergies. Mitig Adapt Strat Glob Chang 12(5):855–873
Schneider UA, McCarl BA (2003) Economic potential of biomass based fuels for greenhouse gas emission mitigation. Environ Resour Econ 24:291–312
Schulze E-D, Freibauer A, Matthes FC, Herold A, Frank Wouters G, Höhne N (2006) Kyoto Protocol: analysis of options for further development of commitments for the second commitment period, part "Sinks in the Second Commitment Period". Umweltbundesamt (Federal Department of the Environment), Dessau

Smith P, Martino D, Cai Z, Gwary D, Janzen H, Kumar P, Smith J (2008) Greenhouse gas mitigation in agriculture. Philos Trans R Soc Biol Sci 363(1492):789–813

Smith P, Martino D, Cai Z, Gwary D, Jazen H, Kumar P, Sirotenko O (2007a) Agriculture. In: Metz B, Davidson O, Bosch P, Dave R, Meyer LA (eds) Climate change 2007: mitigation of Climate Change. Contribution of Working Group III to the Fourth Assessment Report of the Intergovernmental Panel on Climate Change. Cambridge University Press, Cambridge, pp 497–540

Smith P, Martino D, Cai Z, Gwary D, Jazen H, Kumar P, Towprayoon S (2007b) Policy and technological constraints to implementation of greenhouse gas mitigation options in agriculture. Agric Ecosyst Environ 118(1–4):6–28

Theesfeld I, Schmidt O, Scheumann W, Speranza CI (2011) Adapting agricultural water governance to climate change: experience from Germany, Spain and California. Discussion paper 6/2011. German Development Institute, Bonn

UN (1992) United Nations framework convention on climate change. FCCC/INFORMAL/84. United Nations (UN), New York. Retrieved 12 November 2011

UN (1998) Kyoto protocol to the United Nations Framework Convention on Climate Change. United Nations (UN), New York. Retrieved 9 October 2011, from http://unfccc.int/resource/docs/convkp/kpeng.pdf

UNFCCC (2011) Background on the UNFCCC: the international response to climate change. Retrieved 14 December 2011, from United Nations Framework Convention on Climate Change: http://unfccc.int/essential_background/items/6031.php

Vermeulen SJ, Aggarwal PK, Ainslie A, Angelone C, Campbell BM, Challinor AJ, Nelson GC (2010) Agriculture, food security and climate change: outlook for knowledge, tools and action. CCAFS Report 3. CGIAR-ESSP Program on Climate Change, Agriculture and Food Security, Copenhagen

Wollenberg E, Christine N (2011) Next steps for climate change mitigation in agriculture. CCAFS Policy Brief no. 2. Climate Change Agriculture and Food Security (CCAFS), Copenhagen. Retrieved 10 September 2011, from http://www.ccafs.cgiar.org/

Chapter 5
Cost and Opportunities from Mitigation and Adaptation in Agriculture

Abstract It has been predicted that even though efforts are made to reduce drastically the GHG emissions, climate change would still occur as it takes long time for climatic processes to respond. Thus, in addition to mitigation, adaptation to climate change equally plays an important role. Agriculture is one of such sectors that remain highly vulnerable to climate change. It is presumed that the least developed countries are the most susceptible ones due to their higher dependency on climate sensitive sector such as agriculture, and is expected to aggravate further the regional disparities. The international climate change regime designed Clean Development Mechanism (CDM) and Reducing Emissions from Deforestation and Forest Degradation (REDD) to help facilitate industrialized countries to meet their emission targets. Developing countries could be a part of it from where they can fulfill their dual purpose of fulfilling own responsibility of achieving sustainable development and at the same time earn income through carbon finance. Organic farming, bio-gas, System of Rice Intensification (SRI) and community forestry are some of the potential areas discussed where developing countries could take such opportunity. However, such tools are often criticized for its complexities in international implementation rule and varied policies of host nations.

Keywords Adaptation • Agriculture • CDM • Climate change • REDD

5.1 Introduction

The significant effects of climate change are felt all over the world through an increase in global average air and ocean temperatures, widespread melting of snow and ice, and rising global average sea level. It has been substantiated that anthropogenic activities are the main source for such higher concentration of greenhouse

K.L. Maharjan and N.P. Joshi, *Climate Change, Agriculture and Rural Livelihoods in Developing Countries*, Advances in Asian Human-Environmental Research,
DOI 10.1007/978-4-431-54343-5_5, © Springer Japan 2013

gases (GHGs). Fossil fuel use and land use change are primarily responsible for increases in carbon dioxide whereas agricultural sector is accountable for higher concentration of methane and nitrous oxide in the atmosphere (IPCC 2007). If the current pace of GHG emissions is to be continued or even raised, the average global temperature will probably increase by 3–7 °F by 2100. According to the IPCC, if temperatures increase beyond the range of 3.5–5.5 °F over the next 100 years, there would be serious impacts of climate change. Even if efforts are made to reduce the GHG emissions back to 2000 levels there will still be a temperature increase of about 1 °F over the next 100 years as it takes a long time for climatic processes to respond (EPA 2009). Therefore, in addition to mitigation, it is equally important to take adaptive measures to the changing atmosphere to reduce vulnerability imposed by climate change.

One of the sectors that remain highly vulnerable to climate change is agriculture. It is vastly sensitive to climate change mainly due to rising temperature, change in precipitation, frequent and severe occurrence of natural calamities and salt-water invasion caused by a rise in sea level (discussed in Chap. 3). It has been predicted that climate change is most likely to be severe for least developed countries in the tropical and subtropical areas mainly due to their economic dependency on climate sensitive sector such as agriculture and fisheries and limited human, institutional and financial capacity to face the effects of climate change, thus making them more vulnerable (Sem 2009).

According to the projections, climate change might exacerbate regional disparities even if global food production might be able to fulfill the growing food requirement. This is due to reduction in crop yields mostly in places located at lower latitudes where many developing countries are situated (Rosenzweig and Tubiello 2007). According to the estimation made for 2050, there will be an increase in global mean temperatures and weather variability, having implications on agricultural productivity worldwide. In addition to adapt the significant adverse impacts of climate change, agriculture also faces the challenge of increasing its productivity in order to feed a growing population. World population is estimated to reach more than nine billion in 2050 and thus will require food production to double from current levels (OECD 2010). Changes in rainfall patterns will likely result in severe water shortages and/or flooding; melting of glaciers also poses threat of flooding and soil erosion; and rising temperatures will cause shifts in crop growing seasons, which will have huge implications on food security. Climate change will exacerbate the pressure on the already vulnerable people, particularly in developing countries because of lack of assets to protect them from and overcome the shocks. The frequency and intensity of heat waves and heavy precipitation events are already evident and other weather extremes like tropical cyclones, floods, droughts and heavy precipitation events are expected to rise even with increase of relatively small average temperature (UNFCCC 2007). There is, therefore, an urgent need for climate-sensitive sectors of the economy such as agriculture to adapt to ongoing changes in the climatic system so as to reduce its vulnerability.

5.2 Adaptation Strategies in Agriculture

Even though some places might benefit from increasing temperature, the global food production is estimated to decline in the face of climate change. In Africa, climate change has affected agricultural production due to loss of land, shorter growing seasons and more uncertainty about planting time. Yield from rain-fed crops has been predicted to be halved by 2020 in some countries and net revenues from crops to be decreased by 90 % by 2100, exacerbating food insecurity situation. The region is already facing diminishing fish stock due to rise in water temperatures. Likewise, in Latin America food insecurity has been projected to increase due to desertification, salinization and erosion; reducing crop yields and livestock productivity. The small island developing states also remain vulnerable to climate change by sea-level rise, inundation, soil salinization, seawater intrusion into freshwater lenses, and a decline in freshwater supply. Increasing sea surface temperature, rising sea level and damage from tropical cyclones has affected fisheries as well. In Asia too, decrease in crop yield in many parts has aggravated the risk of hunger. Though agriculture has been anticipated to expand in northern areas, reduction in soil moisture and evapotranspiration can increase land degradation and desertification (UNFCCC 2007).

Reversing the implication of climate change is a long-term process and not adjusting to the changing setting will have further implications in agricultural productivity. Adaptive measures, thus, should be taken to minimize the sensitivity of a system to changes in climatic conditions or to exploit new opportunities. The farm management practices should be altered to cope with changing climatic conditions. Adaptation is, however, a complex process as climate change has varying effects on different regions. Therefore, the risks and opportunities related to climate change will also vary by region, which is why adaptation strategies must be developed based on these variations. For instance, within the United States there has been as much as 3 °C increase in average temperature in North Dakota and Oregon, whereas in some parts such as Georgia and Mississippi, average temperatures have decreased by as much as 3 °C. This shows how temperature can vary within the country from the global average temperature increase of 0.3–0.6 °C in the past 100 years. Similarly, the global average precipitation levels have increased 1 % during the same period. However, the mean level of precipitation increased by 20 % in the Susquehanna River Basin, northeast USA and New Mexico but in places like California and Wyoming, the mean level of precipitation decreased by as much as 20 %. In China also, during 2006, East and South part of the country was affected by major storms and flooding, whereas Central, Western and North-eastern regions suffered from heat and drought taking thousands of lives and causing US$20 billion in damage (Scheraga and Grambsch 1998).

Asia being the largest continent is expected to have varying forms of climate change. It has been predicted that increase in rainfall during the summer monsoon could increase flood prone areas in East, South and Southeast Asia. Likewise, glacier melting during the wet season will have significant risk of flooding, erosion, mudslides and GLOF in Nepal, Bangladesh, Pakistan and North India, whereas an increase in frequency and duration of severe heat waves and humidity in summer is expected in temperate and tropical Asia (UNFCCC 2007). As with the case in agriculture sector, it is obvious that these climatic variations will have varying effects on yields as well. For instance, New England may benefit from planting corn through an increase in yields but the Great Plains might experience decline in yields (Scheraga and Grambsch 1998). Similarly, in Central and South Asia, crop yields are predicted to fall up to 30 %, thus increasing the risk of hunger in many countries. Therefore, it requires good understanding of current and potential effects of climate change across different populations and geographic regions. Knowing the mechanism by which the impact will occur is also imperative for effective adaptation measures, the lack of which might lead to a solution that is worse than the problem itself. Such as in agriculture, areas suffering from reduced rainfall and higher evaporation should expand irrigation facilities so as to ensure water availability for uninterrupted yield. As for areas affected by sea level rise, reforestation might be a solution for improving coastal protection (UNFCCC 2007).

There are mainly two types of adaptation measures: reactive, which require actions to be taken as a response to climate change or anticipatory, which is taking measures in anticipation of future climate change (Smith and Lenhart 1996). The national communications of developing countries have highlighted agriculture and food security as key vulnerable sectors. Hence, as an instant response to climate change, certain reactive adaptation measures have been proposed. These include erosion control, dam construction for irrigation, changes in fertilizer use and application, introduction of new crops, soil fertility maintenance, changes in planting and harvesting times, switch to different cultivars, educational and outreach programs on conservation and management of soil and water. Likewise, in order to be prepared for future changes, communities ought to take certain measures. Such as developing crop varieties tolerant or resistant to drought, salt, insects and pests or taking precaution through soil water management. Diversification and intensification of food and plantation crops; developing policy measures through tax, incentives/subsidies or free market; developing early warning systems and extensive research and development has been recognized as other anticipatory measures for protecting from future climate change (UNFCCC 2007).

As Smith and Lenhart (1996) have also identified developing new crop types that are more resistant to climate change and enhance seed banks to allow farmers to diversify their production for adaptation of agriculture to climate change. In such case, avoiding monoculture and adoption of wider crop varieties will also help in reducing vulnerability in case of single crop failure. Similarly, subsidies, taxes or community support programs for certain crops should be discouraged so that farmers can change the cropping system that is best suited to the changing climate. The irrigation facility, which ensures continuous water supply for yield, information dissemination on conservation management practices, liberalizing agricultural trade and promoting agricultural drought management through information and incentives has been classified as strategies for agricultural adaptation to climate change.

Agriculture demands about 70 % of the world's freshwater. The further pressure on increase in agricultural production in the face of resource constraint, particularly water, and higher incidence and severity of floods and droughts require both rain-fed and irrigated agriculture to be managed in a more sustainable way for reducing production risks. Similarly, besides the adoption of new crop or crop varieties, another adaptation measures could be the adoption of animal breeds that are more appropriate to future climate conditions. Government policy can also play an important role in adaptation of agriculture to climate change through reforms in agricultural policies. The information should be given on a large scale to make the public aware of the climate change impact on the agricultural sector and how it can be adapted as well as mitigated with the given cost and benefit. The policy should encourage low GHG emitting production techniques and efficient use of resources through risk management, research and development and market-based approaches. The examples of these may include crop and disaster insurance, research on crop varieties and animal breeds that can better adapt to changing climatic conditions, and incentives for more efficient use of water (OECD 2010).

Climate change will have effect on crop yields, irrigation demand and pest management. Therefore, for successful adaptation strategy in agriculture, it is important to have regional knowledge on crop varieties most suitable for a given climate along with information on planting time, irrigation, fertilizer use, tillage, etc. In a developing country where people lack resources to cope with hazards put forth by climate change and cannot immediately and effectively practice mitigation measures in the short term, people should be provided with the capacity to adapt to change. It requires increasing peoples' ability by warning them about any untoward incident in advance and preparing to deal with vulnerability and uncertainty. Particularly in agriculture, drought-tolerant crop-varieties, mixed cropping, irrigation facilities, research and development, educational and promotional activities, soil-water management, reform in government policies, etc. are important in adaptation process.

5.3 Clean Development Mechanism (CDM) and Reducing Emissions from Deforestation and Forest Degradation (REDD)

5.3.1 Clean Development Mechanism (CDM)

The Kyoto Protocol was established in 1997 as a legal framework to promote international GHGs reduction. It is a protocol to the United Nations Framework Convention on Climate Change (UNFCCC) as a global response to minimize human induced greenhouse gases (GHGs) emission. Under the Kyoto Protocol, industrialized countries are legally bound to bring their emission of GHGs to an average 5 % below 1990's level within the period of 2008–2012. To facilitate this venture, the protocol has established three market-based mechanisms: International Emission Trading; Joint Implementation (JI); and the Clean Development Mechanism (CDM). It is designed to facilitate industrialized countries to meet their emission targets by participating in emission reductions actions at lower costs, either by implementing at their own or in other countries.

Clean Development Mechanism (CDM), which came into force in 2005, is the one in which both developed and developing countries can collaborate for emission reduction. Under this system, entities from Annex I (developed) Parties are to implement emission-reducing projects in non-Annex I (developing) countries, and generate tradable credits in accordance to the volume of emission reduced due to that project. Tradable credits are known as Certified Emission Reduction (CER). They are equivalent to 1 ton of Carbon Dioxide (CO_2) reduction. CER prices are determined by both buyer and seller by evaluating the project based on risk factors involved and prevailing market forces. These kinds of projects help to generate income for the sellers of those credits, and buyers can have some flexibility in meeting their emission reduction or limitation targets. Thus, the CDM's main objectives are to assist developing countries in achieving sustainable development and developed countries in achieving the Kyoto Protocol compliance.

However, there are certain criteria for assessing and approving the project. In order to be a part of the CDM projects; there should be voluntary participation, ratify the Kyoto Protocol, and finally establish a National Authority also called Designated National Authority (DNA) to assess CDM projects. The government should be able to assess the project against their national sustainable development objectives such as poverty reduction, environmental improvement, etc. Consequently, proper facilities for implementation of CDM projects such as publishing investment rules and procedures, identifying desirable projects and how those can be promoted to potential international investors should also be developed. Projects of several categories can be considered for reducing, avoiding or sequestering greenhouse gas emissions such as renewable energy, energy efficiency, fuel switching, co-generation, forestry, transportation, and waste management.

With growing level of greenhouse gases in the atmosphere, an urgent need to combat against climate change has been realized throughout the world. In 2009, at United Nations Climate Change Conference in Copenhagen, political leaders' emphasized their strong political will to urgently combat climate change in accordance with the principle of common but differentiated responsibilities and respective capabilities. Climate financing has been recognized as a key measure to such action for which strong cooperation among nations would be essential. Developed countries have committed themselves to a goal of jointly mobilizing US$100 billion a year by 2020 to address the needs of developing countries for mitigation actions and transparency on implementation of projects. The Advisory Group has stressed the significance of a carbon price in the range of US$20–25 per ton of CO_2 equivalent in 2020 so as to reach US$100 billion per year. It was also estimated to cause flow of US$24–37 billion from developed to developing countries within the next 10 years (Thorne and Rovere 1999).

According to CDM Annual Report 2010, there was a registration of the 2000th project activity (UNFCCC 2010a). Of the total 2,453 registered projects, the highest share is in Asia and Pacific region (77.95 %) followed by Latin America and the Caribbean (19.57 %), whereas Africa (1.96 %) and other (0.53 %) regions are still far behind in active involvement in the global CDM market. The United Kingdom of Great Britain and Northern Ireland, Japan, Switzerland, Italy, France, Spain, Germany, Sweden and the Netherlands are prominent investors in this market, whereas China, India, Republic of Korea, Brazil and Mexico are the major host parties who issue CERs.

CDM encourages significant investment, which creates jobs and thus, helps reduce poverty in developing countries; in addition to facilitating in achieving sustainable development. It helps bring innovation among entrepreneurs to look for more sustainable practices. In the process of financing and implementing the projects, host countries can get introduced to new technologies and knowledge-base as well. The amount generated by way of selling CERs can be used to subsidize/pay for the project implementation or utilized in other activities. The enormity with which carbon financing is growing and is expected to grow, and the prospect of contributing in nation's sustainable development; developing countries should grab the opportunities of the global investment under CDM by setting up a proper arrangement to facilitate and regulate the CDM market in their countries.

5.3.2 Reducing Emissions from Deforestation and Forest Degradation (REDD)

Nearly 20 % of global greenhouse gases are emitted through deforestation and forest degradation, which is more than emission from the entire global transportation sector and second only to the energy sector (UN-REDD 2009). Forests, which are crucial for providing a wide range of socio-economic and environmental benefits such as contribution to rural livelihood and biodiversity conservation, also has the capacity to store a significant amount of global carbon stock. The total carbon content of global forest ecosystem has been estimated to be at 638 Gt for 2005, which is more than the amount of carbon in the entire atmosphere. However, the deforestation rate has been estimated to be around 13 million hectares per year, causing a net loss of about 7.3 million hectares per year from 2000 to 2005 (UNFCCC 2008).

Deforestation reduction has been deemed to be the single largest way to immediately reduce carbon emissions and do it in a cost-effective way (Holloway and Giandomenico 2009). Still CDM project fails to consider this aspect and their activities relating to forestry are only restricted to reforestation and afforestation for the first commitment period (2008–2012) of the Kyoto Protocol. The mechanism of allowing only reforestation and afforestation under emission trading leads to no monetary benefits from the existing forests that are already contributing in carbon sequestration. This might encourage negative behavior of cutting down the existing forest with the expectation of getting a subsidy for new forest, which will result in major loss of biodiversity and other ecological services provided by the forests. Thus, the large contributions of deforestation to global carbon emission and overlooking this aspect by CDM projects have prompted re-negotiation of climate policy. As a result of this, in 2007 the UNFCCC 13th Conference of the parties (COP13) in Bali, REDD was introduced in the international climate policy regime for the post-2012 period in order to adopt a forest mitigation option with major and most abrupt carbon stock impact.

REDD has been developed to create a financial value for the carbon stored in forests. It is considered as the cheapest way of reducing greenhouse gas emissions within a short time frame (Dutschke et al. 2008). It offers incentives for developing countries or other potential participant to reduce emissions from forested lands and invest in low-carbon action for sustainable development. The term "REDD+" includes the additional role of conservation, sustainable management of forests and enhancement of forest carbon stocks. It is predicted that financial flows for greenhouse gas emission reductions from REDD could reach up to US$30 billion a year. Nonetheless, there are issues such as how to include REDD mechanism in existing national development strategies such as; increase participation of forest communities and indigenous peoples in the designing, monitoring and evaluation of national REDD programs, funding the program, ensuring equal distribution of benefits among all those who manage the forests and monitoring the amount of carbon stored and sequestrated as a result of REDD (FAO 2010).

World Bank has already set aside the budget of US$300 million to assist tropical countries who choose to take part in Forest Carbon Partnership Facility (FCPF); a program launched to manage the forest carbon fund. FCPF will help in the preparation and development of REDD systems in developing countries (Cenamo et al. 2009). However, REDD faces major technical issues. Firstly, whether it should be project-based, which will be easier to implement but poses a possibility of in-country leakage and liability; or national-based, that is flexible to manage the collective forest resources is disputed. When it comes to monitoring deforestation also, cost-effectiveness and examining deforestation and quantifying actual carbon flows with accuracy remains an issue. Next, the process of setting a baseline rate of deforestation for promoting equity and motivating participation also is a matter of concern. The country's historic rate of deforestation as a baseline for reducing national deforestation rate can be biased for those who are already taking initiative to lower their deforestation rate, which will unbalance the incentivizing system. Again, if these baselines once established should allow to be changed over time is also questionable. There is a setback of leakage especially in a project-based approach where conservation of one forest can undermine deforestation at some other place at both national and international levels. Lastly, the continued effort to avoid deforestation remains highly doubtable especially in case of weak governance structures and land-tenure system, which are common problems in countries with rapid deforestation (Davis 2008).

The Brazilian government has already prepared a REDD strategy, which is expected to double income of 200,000 rural forest-based families and reduced fire based cost to society (such as respiratory illness, deaths, agricultural and forestry damages) of US$10–80 million per year. Besides biodiversity conservation provides substantial non-monetized benefits as well. The Brazilian REDD program is expected to have opportunity costs of the reduced emissions to be approximately US$18 billion. Even if the REDD program cannot compensate at a level equal to the opportunity cost, it can be justified through the substantial benefits that the Brazilian society would be able to get (Nepstad et al. 2007). Furthermore, REDD has been estimated to generate US$200 million per year in Vietnam. Indonesia, Papua New Guinea and Vietnam were the original pilot countries in the Asia and Pacific region but countries like Bangladesh, Bhutan, Cambodia, Nepal, Sri Lanka, Solomon Islands and the Philippines have also joined the United Nations Collaborative program on Reducing Emissions from Deforestation and Forest Degradation in Developing Countries (The UN-REDD program) as observers. The UN-REDD program assists them in preparation for REDD readiness to participate in a future REDD mechanism. It requires countries to prepare a REDD strategy that can be incorporated in existing national policies and strategies, and incorporation with indigenous peoples and other forest-dependent communities (FAO 2010).

The demand for carbon finance markets like CDM is growing significantly and is only likely to get higher after 2012 with the potential start of REDD. CDM can contribute to developing countries' sustainable development goals through technology transfer and financial resources, energy production by developing renewable energy, increasing energy efficiency and conservation, reforestation/afforestation, poverty alleviation through income and employment generation and local environmental benefits. On the other hand, REDD can also generate substantial income in addition to reducing climate change, biodiversity conservation and socio-economic development. However, the complexities and details involved in implementation of carbon forestry regarding the issues of technical know-how; equity, control and power in case of community forestry, human resource development and other amendments in policies remains difficult. Nevertheless, the potential participants should learn from those countries that are already taking steps in carbon forestry and incorporate those in their national development agenda.

5.4 Agriculture Related CDM

Agriculture is both influenced by and adds to the global climate change. A good agricultural practice that minimizes the pressure on climate change through low GHGs emissions must be adopted, which at the same time adjust to the fast changing climate. This will definitely continue even if we take measures to control the change (EPA 2009). Currently agriculture accounts for 12 % of global GHGs and it is escalating with further deforestation, tilling of pasture and soil degradation; exposing to higher amount of carbon release. Likewise, rising temperature, decreasing water availability and extreme weather events such as droughts and floods have increased agricultural vulnerability, which will have profound effects on food security (ITC 2007). Given the significant effect agriculture has had on global climate change, one of the ways to reduce emission from this sector is to incorporate it in the carbon trading market such as Clean Development Mechanism (CDM).

According to the estimation cited in Tiwari (2000), there is 1220,000 million tons of the global stock of soil organic carbon (SOC) mass in the upper 1 m layer. If the 75 % of the estimated historic loss of 50–100,000 million tons of soil carbon could be captured, it would be about 40–70,000 million tons or 3,000 million tons C/year, equal to 12–25 years of atmospheric increase in carbon. In terms of cost also carbon sequestration through agriculture sector is cheaper which is estimated between US$10–25 per ton compared to US$13–26.0 per ton and from US$200–250 per ton in forestry and industrial sectors, respectively (Tiwari 2000). Thus, with respect to cost and the enormous amount of carbon reduction potential, agriculture sector offers an attractive option for climate change mitigation.

Conversely, the new climate agreement is not able to judge on good or bad agricultural practice and whether soil carbon sequestration should be included in carbon trading. Carbon trading so far has not been able to prevent emissions from fossil fuel burning in the North and it has been found that Clean Development Mechanism (CDM) credits are being used to subsidize some of the most polluting industries in the South. It was found in 2008, even though 4.9 billion tons of carbon dioxide equivalent (CO_2-eq) emission reductions were traded on global carbon markets, and carbon trading increased by 83 % within a period of just 1 year, it has not led to emission reduction. Pig farms and oil palm plantations are a few of the projects under agriculture related CDM; even so they are debated for their contribution in biodiversity destruction and soil and water pollution (Paul et al. 2009). Nonetheless, Copenhagen negotiation is also attempting to include soil carbon sequestration, which many governments in North have also approved

There are numerous ways through which to minimize emission from agricultural sector, some of which are explained below.

5.4.1 Organic Farming

Organic farming is known to be a sustainable food production system that relies on low external input and uses locally available resources for producing at optimal level. It has a high potential for climate change mitigation as it strictly restricts the energy intensive inputs such as fertilizers and pesticides and improves soil carbon storage capacity. Organic farming is also known to emit less methane from the soil compared to mainstream agriculture. On the adaptation side also it is deemed to be more resistible as it can perform well in case of natural calamities such as floods and droughts and reduces soil erosion and enhances fertility. If organic farming could be included in a carbon financing mechanism such as CDM, it could largely benefit poor and small scale farmers who could get additional incentives for shifting to organic farming. Since organic farming is already obliged to practicing internal control system, the substantial additional transaction cost to monitor carbon changes is also reduced. The certification process in organic practice will contribute to a large extent to meet the standards of CDM.

On the contrary, incorporating organic farming in CDM involves complexities as well. The Kyoto Protocol and World Wide Fund for Nature (WWF) both consider agricultural CO_2 sequestration ineligible for carbon credits as it is presumed to be temporary in nature—a change in the landscape will lead to a release of substantial amount of CO_2 in the atmosphere. However, it is also argued that due to buildup of organic matter; quitting organic farming might not lead to faster release of CO_2 than cutting a tree. Even so, it will not be a benefit for a long-term period. In addition, it involves several effects and certification of emissions reduction will require multiple methodologies and, thus, might be expensive (Tennigkeit et al. 2005). Besides the complexities in international negotiations, developing countries face internal difficulties as well. Unlike in major organic countries, subsidies provided by the government, which is also a major reason for success of organic farming in such countries, is lacking in developing countries, such as India and Nepal.

In India, the government is so far able to establish National Standards for Organic Production (NSOP) and approved four accreditation agencies (all government bodies), which are linked to limited crops. A buffer for small scale farmers in case of failure or at least in the initial phase, market development, reducing costs and simplifying the certification process, awareness and research on site specific crop have been identified as steps to stimulate organic sector (Narayanan 2005). For an agriculture backed economy like Nepal, which is also highly vulnerable to climate change, organic farming provides immense potentiality. The excessive and unbalanced use of agro-chemicals has led to decline in soil fertility and factor productivity, loss of indigenous crop varieties, health hazards and less economic return (Bhatta et al. 2008). This is mainly contributed by government supported fertilizer-based agriculture. Realizing the potential implication of organic farming in Nepalese economy, the government endorsed national standard and guideline for organic farming (NPG 2011). In countries where the government is just starting to gear up to include organic farming into the mainstream agriculture, setting it up for CDM project will be much more challenging.

For organic farming to be deemed for CDM projects, crop specific methodologies and regional or national baselines should be established along with appropriate technologies to monitor emissions and sequestration. Through carbon credits, a certain portion of certification costs can also be covered, if this process is to be followed. Again on the other side, since the CERs will be paid after several years, it will be difficult for small scale farmers to afford the initial expenses.

In developing countries, due to high transaction costs and complex processes CDM projects are not easily accessible to small farmers (Lasco et al. 2011). Overall, organic farming is a high-benefit, low-cost CO_2 reduction system and should be included in the carbon trading scheme. In general, developing countries might face barriers in relation to capacity development, finance and manage CDM projects, economies of scale, and other technical capacities. International carbon finance market along with national government policies should be favorable for farmers to adopt and continue with organic practice, at least for a reasonable time period.

5.4.2 *Bio-Gas*

Biogas is a clean, renewable energy that is produced from organic waste materials; used either to burn or generate energy. It is a substitute for firewood or fossil fuels and contributes in global GHGs emissions reduction and is also known to be less time consuming. The main emission reduction takes place due to change in manure management, switching fuels from fossil fuels and non-renewable biomass for energy generation and substituting energy intensive chemical fertilizer with bio-slurry. The animal manure, which was previously kept idle in an open area as a waste is fermented in the biogas digester, and hence, emission of methane is avoided.

Biogas is gaining momentum in developing countries like Nepal and Vietnam where one of the largest biogas programs of the Netherlands Development Organization (SNV) was implemented. In Nepal, heavy dependence on forest resources has resulted in environmental damage such as deforestation, soil degradation and erosion. About 95 % of many rural areas' energy demand is depended on biomass, while some districts even close to 99 %. Among the forms of biomass consumed, fuel wood forms the highest share, which is 90 %. The 95 % of this form of energy is consumed in the rural domestic sector. The remaining proportion of biomass is composed of 90 % fuel wood, 6.5 % dung and remaining comprises of agricultural residues such as corn stalks and chaff from rice (Bajgain and Shakya 2005). The introduction of biogas has provided with a simple, reliable and accessible energy source. The fuel (dung) required to feed the digesters comes from animals and when these digesters are connected to toilets, it improves sanitation. Biogas, in addition to being a cost-efficient source of energy, also reduces the amount of indoor pollution and saves time. Further its by-product, the digested slurry can be used to fertilize the soil and enhance crop yields. At present there are 140,000 rural Nepali households who use biogas for cooking. These installed plants save 400,000 tons of firewood and 800,000 L of kerosene, thus preventing 700,000 tons of greenhouse gases from emitting into the atmosphere, which is estimated to secure about US$3.5 million per year through the CDM. This has been one of the renowned projects, which provided the model for sustainable energy use along with improving rural livelihood at a lower cost. In Vietnam, there were 25,000 biogas plants completed in 2006 and it has been estimated to reduce GHG emission by 75,000 tons (Teune 2007). Installation of biogas has contributed significantly in poverty reduction along with providing clean and affordable energy source.

With the immense ability to reduce GHGs emission and improve rural livelihood, biogas is one of the promising sectors for inclusion in the CDM project. Developing countries can capitalize on greenhouse gas emission reduction by selling CERs, which will further generate funds for investment in biogas programs. Since the Kyoto Protocol came into force on February 16, 2005, the CDM projects submitted for validation has been increased noticeably. The number of biogas project reached 516, which constituted 11.6 % of CDM projects. The Biogas project is mainly concentrated in Thailand, India, China, Malaysia and the Philippines (Daniel et al. 2009). The Nepal program has already managed to secure and the Vietnam program is in the process of acquiring their CERs as well. However, the complexities in complying with rules and regulations under the CDM have been identified as main drawback, which requires collecting energy and manure related data from each and every household. It has been assumed that monitoring cost might outpace the benefits, thus demotivating small decentralized energy project implementers from applying for CDM (Teune 2007).

The initial cost of installing a bio-digester has been the main barrier for households, which were solved by introducing a sustainable micro-finance support system or financial incentives (Daniel et al. 2009). Thus, in a situation like this, the compliance cost that comes with international carbon financing will add another burden for the potential implementers. The Programmatic Approach (PoA), which has been designed in 2007 to overcome the high transaction cost, the reason why small scale projects are largely excluded in the CDM portfolio, can provide additional revenues through biogas programs. But the transaction and monitoring cost is presumed to be much greater than carbon revenues under this approach. In conclusion, biogas provides economic, environmental and social benefits and the inclusion of it in the carbon trading market have further added to its benefit. However, the high transaction cost and other complex procedures still remains an issue in developing countries.

5.4.3 System of Rice Intensification (SRI)

The System of Rice Intensification (SRI) is a method for growing more rice with fewer inputs through management of plants, soil, water and nutrients. It requires careful planting of young seedlings of about 8–12 days old, singly and with a wide spacing of 25 cm or more to support greater root growth. The soil under this method must be kept moist but well-drained and well-aerated and adding compost or other organic material will help in nurturing soil microbial activity (SRI-Rice 2012). In addition, chemical fertilizers can also be applied for nutrient amendments, and frequent weeding (usually two to three times) during the growing season is also recommended.

The traditional method of rice cultivation keeps fields flooded for extended time periods during the growing season, which is known to emit methane due to decomposition of organic matter present in the soil. Such methane emission from rice fields accounts for 11 % of global methane emissions (CDM 2004). Thus, SRI has been considered for CDM project, which is expected to cause less emission. To analyze the emission reduction, the methane emissions from rice fields where cultivation is done by flooding fields for extended period of time during the growing season and where the SRI project (especially SRI water management) is practiced, is compared. The baseline scenario is selected based on conceivable baseline scenarios and likewise alternative baselines are also evaluated based on national circumstances, policies or regulations and agricultural practices leading to reduced methane emission. A gas sampling is then taken with a statistically significant number of samples to see the amount of emission reduced. The methodology does not measure any leakage issues and emission reductions are, thus, calculated as the net of baseline and project emissions. The primary criteria for SRI water management requires moist but well drained and aerated soil conditions as it is the most relevant practice for the avoidance of methane emissions. To make the process of CER payment simple, it has been proposed to pay the project developer who enters into a contractual agreement in advance with individual farmers with different landholding sizes that practice SRI in the project area and utilizes CER revenue to promote SRI in other areas (CDM 2005).

In addition to mitigation potential, SRI practices also reduce water use in rice cultivation by about 25–50 %, thus making suitable in areas with water scarcity. It also stimulates rice plants to grow larger and deeper roots that make them resistant to drought and other kinds of water stress with increased production. SRI benefits have now been validated in 42 countries of Asia, Africa, and Latin America. It has been reported to have increased yield (50–100 % or more), reduced seed requirements (up to 90 %) and water savings (50 % or more). Reductions in pests, diseases, grain shattering, unfilled grains and lodging have also been claimed by many SRI users (SRI-Rice 2012).

The CDM project for SRI implementation is in the process of implementation in Morang district of Nepal. However, technological and common practice barriers prevent the proposed project implementation. An emission reduction of 1,067,202 tons CO_2-eq is expected to be achieved during the first 7 years of the renewable crediting period and revenue from carbon sales has been the most attractive component of project implementation, without which it will not be economically profitable (CDM 2004). SRI is a technology that can contribute towards mitigation and adaptation to climate change along with other benefits such as improved yield and lower production cost. Given its multi-purpose, a proper implementation strategy should be developed that can be replicated throughout the world for wider adaptation.

5.4.4 Community Forestry

Community forestry is a sustainable management of the forest resource which also includes benefit sharing, mainly through involvement of local people. The positivity of including Reduced Emissions from Deforestation and forest Degradation in developing countries (REDD) as a policy under the United National Framework Convention for Climate Change (UNFCCC) might influence many developing countries to implement forest strategy of their own that will provide a source of funding to help reduce their deforestation rates either by strengthening their capacity to enforce forest law or directly paying forest users for sustainable management. Deforestation is one of the primary causes of GHG emission. Tropical countries alone are responsible for 20–25 % of total global emissions (Skutsch and Ba 2010). Under the REDD mechanism, based on previous rate of deforestation, a reference level will be established for a participatory country; depending on which achievements to reduce deforestation rate would be decided and payment for reducing each ton of carbon stock going into the atmosphere due to program implementation would be done consequently.

Although the REDD might provide a good platform for implementing good governance in the forest sector, it is not free from criticism. There are many challenges that may come while implementing global carbon market such as REDD in developing countries; for instance, influence and benefit of privileged few, international professionals' domination in deciding the amount and value of carbon, uncertainty of carbon markets, nationalization of carbon revenues and complex methodologies. The strong technical, and institutional capabilities that are required for developing a baseline scenario for carbon monitoring in a way that creates a win-win situation for the stakeholders involved is the critical methodological issues related to REDD. Moreover, even though REDD will contribute in carbon emission reduction; it is mainly favorable for industrialized countries as the traditional use of the forest for livelihood might be barred for locals through legally binding contracts. Thus, there is cynicism of its workability in countries like Nepal where forests provide an important part of rural livelihood system (Ojha 2009).

In the process of reducing global carbon emission, REDD should also ensure rights and involvement of local communities, indigenous people and forest dependent poor households from the very beginning of the decision making process. There should also be strong institutional and human resource development at national, regional and local level for equal sharing of benefit from program implementation. Also REDD mechanism should be highly context-specific according to the needs of local stakeholders.

The global carbon market is growing and is likely to grow further with the implementation of programs such as REDD. It provides a good opportunity for reaching emission reduction targets for developed countries under the Kyoto Protocol and enhancing sustainable development for developing countries. However, the mechanism to make it work is complex, which demands strong institutional and human capacity. The inclusion of all stakeholders involved and how benefits can be distributed remains challengeable. Overall, carbon trading mechanism like CDM and REDD provides a good platform for environmental sustainability and at the same time provides a source of income through Certified Emission Reduction (CER) credits.

5.5 Economics of Agriculture Related CDM

Agriculture and land use changes are the major contributors to the global Greenhouse Gas emissions but they also provide abundant mitigation options. Agriculture is the highest emitter of N_2O and CH_4 gases. The emissions of N_2O mainly come from the nitrogen content in the soil due to synthetic and organic nitrogen fertilizers, animal shelter and manure management. Methane emissions, on the other hand, mainly comes from enteric fermentation by ruminants, anaerobic turnover in rice paddies, manure handling and soil compaction due to the use of heavy machineries; whereas biomass burning is responsible for both CH_4 and N_2O emissions. By changing the agricultural practice, there are plenty of ways, which minimize emissions from this sector. As identified by the Fourth Assessment Report of the Intergovernmental Panel on Climate Change, there are basically four ways to reduce GHG emissions from agricultural sector: crop rotations and farming system design; nutrient and manure management; livestock management, pasture and fodder supply improvement; fertile soil maintenance and restoration of degraded land (FAO 2009). Besides the mitigation factor, these practices also have a positive impact on agricultural productivity. However, if included in carbon emissions trading market such as Clean Development Mechanism (CDM) under the Kyoto Protocol of the United Nations Framework Convention on Climate Change (UNFCCC), it could also provide additional income for developing countries.

CDM market has been growing at a remarkably high rate. The annual investment in registered CDM projects increased from US$40 million in 2004 to over US$140 billion in mid-2011 (UNFCCC 2010b). The share of agriculture related projects is also quite significant. By defining agricultural projects as those using agricultural residues, outputs or agricultural processes that directly or indirectly reduce greenhouse gases, Larson et al. (2011) identified 17 % of CDM projects to be classified as agricultural and land-use forestry projects out of 5,824, which were active as of December 1, 2010. This shows a significant share of agriculture projects, next to hydro-power and alternative energy categories. Even so the small scale nature of these projects makes the mitigation impact relatively smaller as well. Countries like China, India, Brazil, Mexico and Malaysia account for the largest number (79 %) of agricultural projects. Whereas, in case of Africa, even though the agriculture projects are relatively less, it accounts for a significant part of the few forestry projects.

Besides the mitigation potential, CDM has also been designed to achieve sustainable development and improving livelihoods in the area where the project is implemented through technology transfer and availability of services. The UNFCCC has compiled some of the prototypes, which have been able to successfully achieve the expectations. One of such project is in the Kolar District, Karnataka, India. The Bagepalli CDM Biogas Project has replaced inefficient wood-burning mud stoves, which were traditionally used for cooking. This has served in reducing demand for firewood collection, which was resulting in widespread deforestation, and also utilizing animal dung to produce methane gas. The micro-finance support from CER forward funding helped poor households switch to the cleaner technology, thus eliminating time spent collecting fuel wood, improving indoor air quality, and creating jobs and capacity building to use the technology. Moreover, after the project cost is fully recovered, the CDM revenues will be directly transferred to the participating women. Similarly, the Bagepalli CDM Biogas Project has been successful in managing waste by way of composting in Gianyar, Bali, Indonesia. Doing so has turned the once polluted landfill into a clean area and the organic fertilizer so produced acts as a substitute for imported chemical fertilizers. There are numerous other benefits such as additional income from selling reusable waste to recyclers, local employment generation, enhancing knowledge on the environment, climate change and sustainable energy through training and education (UNFCCC 2010c).

However, there are some complexities in the project implementation, which have been evolved to make the process simpler. The methodologies for assessing mitigation amount of a project can either be the one already in place for a specific purpose or can be modified or introduce new ones, both of which must be approved by an expert committee. Nevertheless, usually more than one methodology is used in a project. For instance, a project implementer who uses biodegradable wastes to generate electricity may use one methodology to demonstrate conversion of methane to less harmful carbon dioxide when methane is burned and another one to measure the benefits of switching from fossil fuel based to renewable based electricity generation. The CDM board, for reducing fixed cost of bringing a project to them and monitoring costs, differentiates between small and large scale projects. CDM projects are difficult to implement mainly due to varied national policies and complexities in international implementation rules (Larson et al. 2011).

Even though the CDM has had an impact in lowering GHG emissions to some extent, it has not been able to fulfill its potential of agricultural mitigation aspects such as carbon sequestration in agricultural soils, which has the highest mitigation potential (Smith et al. 2007). The primary reason for making soil sequestration of CO_2 ineligible for carbon credits under the CDM mechanism is that agricultural CO_2 sinks are considered temporary in nature. A change in land use releases large amount of CO_2 back into the atmosphere either due to human activity such as logging or natural events such as forest fire or disease. Thus, due to such reasons only afforestation or reforestation projects and reducing non-CO_2 gases are included but conservation tillage and restoration of degraded soils are considered ineligible for carbon credit under CDM (Rosegrant et al. 2008; Larson et al. 2011). Although soil carbon sequestration projects has been gaining support outside CDM in emerging markets like Canada and United States through carbon offset credits for no-till practices. In addition to this, the Chicago Climate Exchange also includes conversion of cropland to grasslands in a market trading mechanism (Smith et al. 2007).

References

Bajgain S, Shakya I (2005) The Nepal biogas support program: a successful model of public private partnership for rural household energy supply. Ministry of Foreign Affairs, Kathmandu/SNV-Netherlands Development Organisation, The Netherlands/Biogas Sector Partnership, Nepal

Bhatta GD, Doppler W, KC KB (2008) Problems and potentials of organic agriculture development in Nepal. In: Proceedings of international conference on the Competition for Resources in a Changing World. New Drive for Rural Development. Institute of Agricultural Economics and Social Sciences in the Tropics and Subtropics, Hohenheim, p 256

CDM (2004) Clean Development Mechanism: project design document form (CDM-PDD). United Nations Framework Convention on Climate Change, Bonn

CDM (2005) Clean Development Mechanism: proposed new methodology: baseline (CDM-NMB). United Nations Framework Convention on Climate Change, Bonn

Cenamo MC, Pavan MN, Campos MT, Barros AC, Carvalho F (2009) Casebook of REDD projects in Latin America. Institute for Conservation and Sustainable Development of Amazonas, The Nature Conservancy, Manaus

Daniel B, Ksenia B, Martin B, Elke F, Zoran K (2009) Mini biogas plants for households. GFA ENVEST, Germany

Davis C (2008) Protecting forests to save the climate: REDD challenges and opportunities. Retrieved 11 November 2011, from EarthTrends: Environmental Information: http://earthtrends.wri.org/updates/node/303

Dutschke M, Wertz-Kanounnikoff S, Peskett L, Luttrell C, Streck C, Brown J (2008) How do we match country needs with financing sources? In: Arild A, Angelsen A (eds) Moving ahead with REDD: issues, options and implications. Center for International Forestry Research, Bogor, pp 41–52

EPA (2009) About global warming and climate change: back to basics. United States Environmental Protection Agency, USA. Retrieved 17 September 2011, from http://www.epa.gov/climatechange/downloads/Climate_Basics.pdf

FAO (2009) Low greenhouse gas agriculture: mitigation and adaptation potential of sustainable farming systems. Food and Agriculture Organizations of the United Nations, Rome

FAO (2010) Implementation of REDD and what it means for forestry and agriculture in the region. In: Thirtieth FAO Regional conference for Asia and the Pacific. Food and Agriculture Organization of the United Nations, Gyeongju, p 6

Holloway V, Giandomenico E (2009) The history of REDD policy. Carbon Planet, Adelaide

IPCC (2007) Climate change 2007: the physical science basis. Contribution of Working Group I to the Fourth Assessment Report of the Intergovernmental Panel on Climate Change. Cambridge University Press, Cambridge

ITC (2007) Organic farming and climate change. International Trade Centre, Geneva

Larson DF, Dinar A, Frisbie JA (2011) Agriculture and the clean development mechanism. The World Bank, USA

Lasco RD, Habito CM, Delfino RJ, Pulhin FB, Concepcion RN (2011) Climate change adaptation for smallholder farmers in Southeast Asia. World Agroforestry Centre, Philippines

Narayanan DS (2005) Organic farming in India: relevance, problems and constraints. National Bank for Agriculture & Rural Development, Department of Economic Analysis & Research, Mumbai

Nepstad D, Soares-Filho B, Merry F, Moutinho P, Rodrigues HO, Bowman M, Rivero S (2007) The costs and benefits of reducing carbon emissions from deforestation and forest degradation in the Brazilian Amazon. The Woods Hole Research Center, Bali

NPG (2011) Proceedings of the national policy dialogue workshop on organic agriculture in Nepal. Nepal Permaculture Group, Kathmandu

OECD (2010) Agriculture and climate change: impacts, mitigation and adaptation. Retrieved 22 September 2011, from OECD Agriculture Ministerial Meeting: http://www.oecd.org/document/18/0,3746,en_21571361_43893445_44437010_1_1_1_1,00.html

Ojha HR (2009) Climate change, forestry and carbon financing in Nepal: editorial introduction. J For Livelihood 8:1–3

Paul H, Ernsting A, Semino S, Gura S, Lorch A (2009) Agriculture and climate change—real problems, false solutions. The Bonn Climate Change Talks, Bonn

Rosegrant MW, Ewing M, Yohe G, Ian Burton SH, Valmonte-Santos R (2008) Climate change and agriculture: threats and opportunities. German Technical Cooperation (GTZ), Eschborn

Rosenzweig C, Tubiello FN (2007) Adaptation and mitigation strategies in agriculture: an analysis of potential synergies. Mitig Adapt Strat Glob Chang 12(5):855–873

Scheraga JD, Grambsch AE (1998) Risks, opportunities, and adaptation to climate change. Clim Res 10:85–95

Sem DG (2009) The impact of climate change on the development prospects of the least developed countries and small island developing states. United Nations Office of the High Representative for the Least Developed Countries, Landlocked Developing Countries and Small Island Developing States (UN-OHRLLS), New York

Skutsch MM, Ba L (2010) Crediting carbon in dry forests: the potential for community forest management in West Africa. For Policy Econ 12(4):264–270

Smith JB, Lenhart SS (1996) Climate change adaptation policy options. Clim Res 6:193–201

Smith P, Martino D, Cai Z, Gwary D, Jazen H, Kumar P, Sirotenko O (2007) Agriculture. In: Metz B, Davidson O, Bosch P, Dave R, Mayers LA (eds) Climate change 2007: mitigation of climate change. Contribution of Working Group III to the Fourth Assessment Report of the Intergovernmental Panel on Climate Change. Cambridge University Press, Cambridge, pp 497–540

SRI-Rice (2012) About the SRI international network and resources center (SRI-RICE). Retrieved 10 January 2012, from SRI-Rice: System of Rice Intensification: http://sri.ciifad.cornell.edu/aboutsri/aboutus/index.html

Tennigkeit T, Bodnar F, Dutschke M (2005) Mitigation and adaptation potential of organic farming. UNFCCC COP 11/MOP 1. United Nations Framework Convention on Climate Change, Montreal

Teune B (2007) The biogas programme in Vietnam; amazing results in poverty reduction and economic development. SNV Vietnam, Biogas Programme Division, Hanoi

Thorne S, Rovere DE (1999) Criteria and indicators for appraising clean development mechanism (CDM) projects. Helio International, Paris

Tiwari DN (2000) Towards a framework for the implementation of the Clean Development Mechanism in the agricultural sector of developing countries. In: Two essays on climate change and agriculture: A developing country perspective (FAO Economic and Social Development Paper 145). Food and Agriculture Organization of the United Nations, Rome, pp 35–85

UNFCCC (2007) Climate change: impacts, vulnerabilities and adaptation in developing countries. United Nations Framework Convention on Climate Change (UNFCCC), Bonn. Retrieved 5 September 2011, from http://unfccc.int/resource/docs/publications/impacts.pdf

UNFCCC (2008) Forests, climate change, biodiversity and land degradation. Joint Liaison Group of the Rio Conventions (UNFCCC CBD and UNCCD), Bonn

UNFCCC (2010a) Clean development mechanism executive board: annual report. United Nations Framework Convention on Climate Change (UNFCCC), Bonn

UNFCCC (2010b) The contribution of the Clean Development Mechanism under the Kyoto Protocol to technology transfer. United Nations Framework Convention on Climate Change (UNFCCC), Bonn

UNFCCC (2010c) CDM development benefits. Retrieved 26 December 2011, from CDM UNFCCC: http://cdm.unfccc.int/about/ccb/index.html

UN-REDD (2009) About REDD+. Retrieved 25 September 2011, from UN-REDD Programme: http://www.un-redd.org/AboutREDD/tabid/582/Default.aspx

Chapter 6
Methodologies to Assess the Impact of Climate Change in Agriculture

Abstract This chapter discusses different methodologies used to assess the impact of climate change on agricultural production. Crop bio-physical simulation models and regression models are the methodologies widely used to estimate the effects of environmental changes (climate change) on agricultural production through agricultural productivity. Most studies on the possible impact of climate change on crop yields used mainly indirect crop simulation models that make use of crop biophysical simulation. Predictions of the yield changes in response to changes in climate variables, from regression models based on historical climatic and yield data for specific crops are also widely used. This can be done through application of production function. Each of the methodologies used to assess the impact of global climate change on agricultural production is described in this section.

Keywords Crop bio-physical simulation • Indirect crop simulation • Prediction • Production function • Regression

6.1 Crop Biophysical Simulations

Most studies on the possible impact of climate change on crop yields used mainly indirect crop simulation models that make use of crop biophysical simulation (Porter and Semenov 2005). The important thing we have to understand is that these crop biophysical simulation models are different for different crops. For instance, SOYGRO model is specific for soybean, whereas CERES model is for Maize, WTGROWS for wheat and SUCROS for cotton. All these models are process oriented that consider different phenological stages, crop-soil carbon and nitrogen balance, and soil-water balance under different climate related treatments.

K.L. Maharjan and N.P. Joshi, *Climate Change, Agriculture and Rural Livelihoods in Developing Countries*, Advances in Asian Human-Environmental Research,
DOI 10.1007/978-4-431-54343-5_6, © Springer Japan 2013

Based on the assessment of responses under different treatments, future impacts of climate change on agriculture is assessed. In doing so, the outcome of climate model provides the crop biophysical simulation model with necessary inputs on climate variables like daily totals of shortwave radiation, rainfall, daily maximum temperature and minimum temperatures. Thus, based on these data from climate model, the crop biophysical simulation model calculates daily updated values of plant biomass, evaporation, and transpiration as well as the remaining water, carbon and nitrogen in the soil profile as well as a total grain yield under different scenario considered for the study (Takle and Pan 2005). Besides, such crop biophysical simulation models can also be coupled with the models like the CENTURY that capture dynamics within soil organic matter.

Since the initial assessment of response to different controlled variables is based on the controlled experiment, such models have limitations of isolation from the variety and variability of factors and conditions that affect production at the field condition (Adams et al. 1998). Thus, these types of models have limitation on properly understanding the effects of a wide range of variables associated with global warming (Schlenker and Roberts 2008). In addition, though it is unequivocal that global warming is inevitable in the coming century, even if emissions of greenhouse gases is stabilized at current level, there exists debate and uncertainty on the extent of warming as well as other related changes (IPCC 2007; Rosegrant et al. 2008). Similarly, due to huge cost involved in installing the experiment setup, application of such models in the case of developing countries is very limited.

6.2 Ricardian Approach

The production-function approaches that are commonly employed in crop biophysical simulation models is the traditional approach of estimating the impact of climate change on yield of specific crop using either empirical or experimental production functions. Such approaches take an underlying production function and estimates impacts by varying one or a few climatic variables as independent variables such as, temperature, precipitation, and/or CO_2 level. However, these approaches are not able to capture a full adjustment (adaptation) made by farmer in response to changing environmental conditions. Most studies employing such approaches simply calculate the impact of changing temperature on farm yields or allowed limited changes on fertilizer application or irrigation. Thus, there always remains a possibility of having inherent bias that leads to overestimation of damage. The inherent bias is also termed as the "dumb farmer scenario" (Mendelsohn et al. 1994). Therefore, by not permitting a complete range of adjustments, previous studies using such approaches have overestimated the impact of climate change on agricultural production through adverse impacts on crop yields.

To correct the inherent bias in the production-function approach and thereby overestimation of damage in agricultural production Mendelsohn et al. (1994) developed Ricardian approach. This approach is based on the Ricardian Theory of

Rent, which states that the value of land in terms of rent is determined by the quality of that particular piece of land. Therefore, instead of studying yields of specific crops the Ricardian approach uses economic data on the value of land and examines how climate in different places affects the net rent or value of farmland. The net rent or value of farmland is assessed by directly measuring farm prices or revenues per unit of land. Such assessments take into account of the direct impacts of climate on yields of different crops as well as the indirect substitution of different inputs, an introduction of different activities, and other potential adaptations to different climates. Therefore, key model assumptions for the Ricardian approach are as follows (Mendelsohn et al. 1994; Wang et al. 2009):

(i) Each farmer wishes to maximize income subject to the exogenous conditions of his/her farm. Therefore, adaptation takes place by all means including the adoption of new crops and farming systems.
(ii) Markets are properly functioning under perfect competition for land, where free entry and exit will ensure that excess profits are driven to zero. Therefore, the land rent will be equal to the net income per unit of land through its best use.
(iii) The land values have attained the long-run equilibrium associated with each region's climate, therefore does not take into account the cost of transition.
(iv) Market prices are unchanged.

Assumptions (iii) and (iv) are usually criticized by the critics of Ricardian approach. Since the approach is a cross-section analysis, it does not account the dynamic transition costs, which usually occur while farm moves between two states. Similarly, by not taking the price changes into account, the approach will overestimate welfare effects (Mendelsohn et al. 1994; Cline 1996). However, since the prices of crops are determined globally, with the expansion of crop production in some parts of the world and the contraction in other parts, the prices of crops from global warming is expected to be small, which matches the assumption of the Ricardian approach. Despite such criticism, which itself are not fatal, it has the advantage of cost effectiveness and therefore is widely used even in the case of several developing countries (Mendelsohn et al. 1994).

Box 6.1 Ricardian Theory of Rent

Rent is that portion of the produce of the earth, which is paid to the landlord for the use of the original and indestructible powers of the soil. It is often, however, confounded with the interest and profit of capital, and, in popular language, the term is applied to whatever is annually paid by a farmer to his landlord. …that a portion only of the money annually to be paid for the improved farm, would be given for the original and indestructible powers of the soil

Source: David Ricardo (1817).

6.2.1 Derivation of Ricardian Approach

As listed in the list of assumptions, Ricardian approach assumes that each farmer wishes to maximize income, subject to the exogenous conditions of his/her farm. Therefore, the farmer chooses the crop and inputs for each unit of land that maximizes annual income (Wang et al. 2009);

$$Max\,\pi = \sum_i P_{qi} Q_i \left(X_i, L_i, K_i, IR_i, C, W, S \right) - \sum_i P_X X_i - \sum_i P_L L_i - \sum_i P_K K_i - \sum_i P_{IR} IR_i. \quad (6.1)$$

Here, π is net annual income, P_{qi} is the market price of crop i, Q_i is the production function for crop i, X_i is a vector of annual inputs such as seeds, fertilizer, and pesticides for each crop i, L_i is a vector of labor for each crop i, K_i is a vector of capital such as tractors and harvesting equipment for each crop i, IR is a vector of irrigation choices for each crop i, C is a vector of climate variables, W is available water for irrigation, S is a vector of soil characteristics, P_X is a vector of prices for the annual inputs, P_L is a vector of prices for each type of labor, P_K is the rental price of capital, and P_{IR} is the annual cost of each type of irrigation system. Therefore, based on the same assumption the net income will be a function of the exogenous variables.

$$\pi^* = f(P_q, C, W, S, P_X, P_L, P_K, P_{IR}) \quad (6.2)$$

Thus, the model can be specified more simply in the following way

$$V = b_0 + b_1 T + b_2 T^2 + b_3 P + b_4 P^2 + b_5 S + b_6 irri \ldots + b_n edu + e \quad (6.3)$$

Where, V (dependent variable) is crop net revenue per unit of land. b_1, b_2 … b_n are coefficients of respective variables. Similarly, T and P represent temperature and precipitation of the season of interest. Square of T and P is introduced in the model to capture the expected nonlinear relationship between net revenue and climate. S represents soil quality, irri represents irrigation coverage and edu represents the education level of either household head or each member of a household that is in the labor force. Several socio-economic and demographic variables can be accommodated in the model. However, due care should be given while considering the independent variables in the model. All independent variables must be important from a theoretical point of view and would not affect the efficiency of model through multicollinearity, or endogenity, or heteroscedasticity or other specification problem.

6.3 Time Series Analysis

Use of past time-series data on yield and climate variables is another important methodology to establish relationship between climate variables and yield of agricultural crops. Such analysis will be crucial in understanding future climate change effect on agriculture (Nicholls 1997). The possibility of capturing the effect of climate trends over the past period on agriculture was first put forth by Nicholls (1997) through his seminal paper published in the *Nature*. The same methodology was applied by Lobell and Asner (2003) to assess the relation between climate variation and crop production in the US, which was published in *Science*. Since then, the methodology has been widely employed to assess the impact of climate variables on yield of several crops in the global scale, country level or regional level (Peng et al. 2004; Lobell et al. 2005; Lobell and Field 2007; Joshi et al. 2011). Regression equation used to assess the effect of climate variables on agricultural crop yield is as follows:

$$\Delta Yield = m + r_y \ Climate + \varepsilon \tag{6.4}$$

Here, $\Delta Yield$ is observed trend in yield, m is the average yield change due to management and other non-climatic factors, $\Delta Climate$ is observed trend in temperature or rainfall, r_y is yield response to climate trend, and ε is residual error. Residual value obtained after detrending of yield and climate variables are fed into the Eq. (6.4). Non-climatic influences such as new cultivars and changes in crop management practices can be removed by detrending all variables and using the residuals to calculate quantitative relationships between variation in climate and yield (Nicholls 1997; Lobell and Asner 2003). Similarly, the regression analysis is done by forcing the intercept to pass through the origin, which helps to avoid trend effects (Nicholls 1997).

In the regression analysis, rather than using annual averages of each climate variables, each of the climate variables used in such models should coincide with the effective growing season of the crop that is considered for the study. For instance, to assess the effect of climate variables on yield of rice in Saijo, Hiroshima, Japan, temperature and rainfall data from July to October, which covers the period from planting to harvesting, should be taken into consideration. However, while assessing the regional effects of climate change on yields of agricultural crops, an effective 'global growing season' for crops based on the contiguous months within the growing seasons for the major growing regions that produce the highest model R^2should be considered (Lobell and Field 2007).

Once the slope or coefficient for each of the independent variables is obtained through regression model, the net effect of climate variables on yield of agricultural crops can be assessed. This can be done by putting the value of coefficients and changes in climate variables in following equation:

$$\Delta Yield = \beta_1 * \Delta T_{max} + \beta_2 * \Delta T_{min} + \beta_3 * \Delta R \tag{6.5}$$

Here, $\Delta Yield$, ΔT_{max}, ΔT_{min}, and ΔR are the trends in the respective variables between the periods under study. For instance, if the study covers the period of 1978–2008, trend in yield is the difference in yield between 2008 and 1978.

This type of analysis assumes that a year-to-year management changes are either uncorrelated with climate or are themselves caused by climate i.e., crop yields respond in the same way to rapid and gradual climate variations. Similarly, these types of models do not attempt to capture details of plant physiology or crop management. However, they do capture the entire ranges of processes by which climate affects yields, including the effects of poorly modeled process. In addition, these models enable a quantitative evaluation of uncertainties. Since, farmers would adapt cropping systems as climate changes, which minimize or possibly reverse the adverse impact of climate change; these types of analyses represent an upper bound on the impact of recent trends. However, some studies have documented that recent trends in management practices are not driven by climate change. In addition, adaptation is expected to lag several years behind climate trends as there is difficulty in distinguishing climate trends from natural variability and the disaggregated nature of farmer decisions. Therefore, under such circumstances, the estimation using this type of regression analysis would be moderate.

Simply, the observed data rather than using the residuals obtained from detrending can also be used while making a time-series analysis. However, such regression analysis should include all the possible variables that affect yields such as irrigation coverage, input use, labor use and etc.

6.4 Panel Data Analysis (Just and Pope Production Function)

The effect of climate variable is not uniform across the regions. Such variation in the impact of climate could not be captured by time-series. Similarly, such analysis also has limitation in capturing the risk in yield variability due to climate variability over the years. The year to year crop yield variability caused by weather conditions is an important source of production risk (Kim and Pang 2009). Therefore, it is not only important to understand how much crop yields change due to climate change but also it is equally important to understand how much yield variability changes. Both aspects are equally important as it could serve as the basis of formulating government agricultural policies and agricultural research program. Stochastic production function could serve in this direction to capture the effect of climate variables on both yield as well as yield variability of the crops. The stochastic production function introduced by Just and Pope (1978, 1979) is successful in capturing both aspects in a single model, and thus is widely adopted. The functional form of the stochastic production function is given as follows [Eq. (6.6)]:

$$y_{it} = f(X_{it}, \alpha) + h(X_{it}, \beta)\varepsilon_{it} \tag{6.6}$$

Where, y_{it} is the yield for region *i* and year *t*; X_{it} is the vector of explanatory variables such as rainfall, maximum temperature, minimum temperature, and time trend; α and β are unknown parameters to be estimated ε_{it} is an error term with a mean of zero and a variance equal to one.

As seen the Eq. (6.6), the stochastic production function has the basic concept of specifying a production function as the sum of two components. The first term in Eq. (6.6), $f(X_{it},\alpha)$ is associated with yield and represents the mean response function where yield is explained by variables given by X_{it}. The second term $h(X_{it}, \beta)$ is associated with the variability of yield and represents the variance function explained by variables given by X_{it}. Here, since the Just–Pope production function does not impose a priori restriction on the risk effects of input, the second function accommodates both increasing as well as decreasing risk effects of inputs on yield. Thus, the sign of *h* indicates whether a climate variable increases or decreases the crop yield variability. The climate variable is said to be risk increasing if it increases the variability of crop yields and said to be risk decreasing if it decrease the variability of crop yield under given uncertainty.

The model can be estimated by a three-stage feasible generalized least squares (FGLS) procedure as well as Maximum Likelihood Estimation (MLE) approach (Just and Pope 1978, 1979). The following three steps are involved while estimating the model.

- First, the regression model on Y_{it} on $f(X_{it},\alpha)$ is estimated by Ordinary Least Squares (OLS). Residuals obtained from the OLS will be an input for the second step.
- Second, the natural log of the squared residuals of the estimated equation is employed to estimate $h^2(X_{it})$. Here, we can regress $\hat{u}_{it}^2$ on $h^2(X_{it}, \beta)$, since $\hat{u}_{it}$ is consistent for u_{it} and $E(u_{it}^2) = h^2(X_{it}, \beta)$.
- Finally the yield response is then estimated as a weighted regression of Y_{it} on $f(X_{it},\alpha)$ with weights $h^2(X_{it}, \beta)$.

Just and Pope Production function that uses FGLS has been employed overwhelmingly by applied production economist and every Just and Pope estimate has used FGLS prior to 1997 (Saha et al. 1997). However, while comparing the MLE and FGLS through the application of Monte Carlo experiment, Saha et al. (1997) shown that MLE is more efficient especially in small samples even when the error distribution departs significantly from normality.

In the case of panel data, before running the model, it is always important to check the nature of data using Panel Unit Root Tests. The basic idea behind this is to check the nonstationary nature of an individual time series data which may cause a spurious regression. Therefore, performing the unit root test to check nonstationarity of panel data is crucial. Such test should be individually applied to each of the potential dependent as well as independent variables. The null hypothesis of a unit root is either rejected or accepted based on the P-value while performing the test. If the P-value is non-significant we have to accept the null-hypothesis that the individual time-series data is nonstationary, which means the time-series data is not eligible for panel data analysis. However, if the P-value is significant, we reject the null hypothesis, thus the variables are stationary as a panel or integrated order zero and is applicable for running the model.

Similarly, it is also necessary to ascertain the correct panel model for the estimation of production functions as panel data models take two alternative forms: random effects and fixed effects (Baltagi 2010). The Hausman test statistics can be used to test the random effects model versus the fixed effects model in order to determine the correct panel data model. This test statistics is distributed asymptotically as chi-squared with explanatory variables' degree of freedom under the null hypothesis that the random effects estimator is consistent and more efficient(Isik and Devadoss 2006). Thus, if P-value is not significant, we have to accept the null hypothesis that random effect estimator is consistent and more efficient.

Only the climate variables together with trend (time variable) could be included in such model analysis taking the regions as dummy (Isik and Devadoss 2006; Kim and Pang 2009). In addition to climate variable, the input variables could be accommodated in such model (Chi-Chung et al. 2004; Carew et al. 2009; Boubacar 2010). It is mostly, land area under cultivation as input variable is used for such estimation. Since the use of input variables reflects the changes in management practice across the region, it is always recommended to include input variables as far as possible.

6.4.1 Steps to Run Just and Pope Stochastic Production Function in STATA

Let us assume the following variables

Yield (Yld), soil (Sl), nitrogen fertilizer (np), precipitation (Prec), temperature (Temp), area (area), dummy for; region1 (dumreg1), region2 (dumreg2), region3 (dumreg3), and region4 (dumreg4).

First Component of the Eq. (6.6)

Testing the significance of the intercept with a Walt test

```
. reg yld Sl np Prec Temp area dumreg1 dumreg2 dumreg3 dumreg4

      Source |       SS       df       MS              Number of obs =     165
-------------+------------------------------           F( 9,     155) =    2.29
       Model |  3.21041727     9  .133767386           Prob > F      =  0.0015
    Residual |   8.1932518   155  .058523227           R-squared     =  0.2815
-------------+------------------------------           Adj R-squared =  0.1584
       Total |  11.4036691   164  .069534568           Root MSE      =  .24192

------------------------------------------------------------------------------
        lyld |      Coef.   Std. Err.      t    P>|t|     [95% Conf. Interval]
-------------+----------------------------------------------------------------
          Sl |   .8090881   1.506747     0.54   0.592    -2.169832    3.788008
          np |  -.8487864   .5626571    -1.51   0.134     -1.96119    .2636168
        Prec |  -.5016811   .1424151    -3.52   0.001    -.7832435   -.2201187
        Temp |   .1734354   .0967151     1.79   0.075    -.0177756    .3646464
        area |   .2410418   .2594448     0.93   0.354    -.2718945    .7539781
     dumreg1 |  -.2178562    .769878    -0.28   0.778    -1.739946    1.304234
     dumreg2 |  -.0316352   .3032999    -0.10   0.917    -.6312753     .568005
     dumreg3 |  -.2614979    .334916    -0.78   0.436    -.9236449     .400649
     dumreg4 |   -.336324   .2639313    -1.27   0.205    -.8581303    .1854822
       _cons |   15.11567   4.504893     3.36   0.001     6.209256    24.02209
------------------------------------------------------------------------------
```

. test _cons
(1) _cons = 0
F(1,155) = 11.26
Prob > F = 0.0010
. * The null that intercept is not significant was rejected.

Test for heteroskedasticiy

```
. reg yld Sl np Prec Temp area
      Source |       SS       df       MS              Number of obs =     165
-------------+------------------------------           F(  5,    159) =    3.78
       Model |  2.24946823     5  .224946823           Prob > F      =  0.0001
    Residual |  9.15420085   159  .059442863           R-squared     =  0.1973
-------------+------------------------------           Adj R-squared =  0.1451
       Total |  11.4036691   164  .069534568           Root MSE      =  .24381

------------------------------------------------------------------------------
        lyld |      Coef.   Std. Err.      t    P>|t|     [95% Conf. Interval]
-------------+----------------------------------------------------------------
          Sl |   .2898455   .1212125     2.39   0.018     .0503916    .5292994
          np |  -.0097866   .2797973    -0.03   0.972    -.5625228    .5429496
        Prec |  -.0729251   .3196656    -0.23   0.820    -.7044207    .5585706
        Temp |   .0967691   .0426348     2.27   0.025     .0125445    .1809936
        area |  -.1520231   .0816022    -1.86   0.064    -.3132272    .0091811
       _cons |   9.513932   2.116144     4.50   0.000     5.333515    13.69435
------------------------------------------------------------------------------
```

. hettest

Breusch-Pagan / Cook-Weisberg test for heteroskedasticity

Ho: Constant variance
Variables: fitted values of yld
chi2(1) = 51.31
Prob > chi2 = 0.0000

Also run test for heteroskedasticity with dummy variables in the equation

```
. reg yld Sl np Prec Temp area dumreg1 dumreg2 dumreg3 dumreg4

      Source |       SS       df       MS              Number of obs =     165
-------------+------------------------------           F( 9,     155) =    2.29
       Model |  3.21041727     9  .133767386           Prob > F      =  0.0015
    Residual |   8.1932518   155  .058523227           R-squared     =  0.2815
-------------+------------------------------           Adj R-squared =  0.1584
       Total |  11.4036691   164  .069534568           Root MSE      =  .24192

------------------------------------------------------------------------------
        lyld |      Coef.   Std. Err.      t    P>|t|     [95% Conf. Interval]
-------------+----------------------------------------------------------------
          Sl |   .8090881   1.506747     0.54   0.592    -2.169832    3.788008
          np |  -.8487864   .5626571    -1.51   0.134     -1.96119    .2636168
        Prec |  -.5016811   .1424151    -3.52   0.001    -.7832435   -.2201187
        Temp |   .1734354   .0967151     1.79   0.075    -.0177756    .3646464
        area |   .2410418   .2594448     0.93   0.354    -.2718945    .7539781
     dumreg1 |  -.2178562    .769878    -0.28   0.778    -1.739946    1.304234
     dumreg2 |  -.0316352   .3032999    -0.10   0.917    -.6312753     .568005
     dumreg3 |  -.2614979    .334916    -0.78   0.436    -.9236449     .400649
     dumreg4 |   -.336324   .2639313    -1.27   0.205    -.8581303    .1854822
       _cons |   15.11567   4.504893     3.36   0.001     6.209256    24.02209
------------------------------------------------------------------------------
```

. hettest

Breusch-Pagan / Cook-Weisberg test for heteroskedasticity
Ho: Constant variance
Variables: fitted values of lyld
chi2(1) = 71.77
Prob > chi2 = 0.0000

*In both cases the null was rejected and therefore there is heteroskedasticity.

Generating residuals (First stage of Just and Pope Stochastic Production Function)

```
. reg yld Sl np Prec Temp area dumreg1 dumreg2 dumreg3 dumreg4

      Source |       SS       df       MS              Number of obs =     165
-------------+------------------------------           F( 9,     155) =    2.29
       Model |  3.21041727     9  .133767386           Prob > F      =  0.0015
    Residual |   8.1932518   155  .058523227           R-squared     =  0.2815
-------------+------------------------------           Adj R-squared =  0.1584
       Total |  11.4036691   164  .069534568           Root MSE      =  .24192

------------------------------------------------------------------------------
        lyld |      Coef.   Std. Err.      t    P>|t|     [95% Conf. Interval]
-------------+----------------------------------------------------------------
          Sl |   .8090881   1.506747     0.54   0.592    -2.169832    3.788008
          np |  -.8487864   .5626571    -1.51   0.134     -1.96119    .2636168
        Prec |  -.5016811   .1424151    -3.52   0.001    -.7832435   -.2201187
        Temp |   .1734354   .0967151     1.79   0.075    -.0177756    .3646464
        area |   .2410418   .2594448     0.93   0.354    -.2718945    .7539781
     dumreg1 |  -.2178562    .769878    -0.28   0.778    -1.739946    1.304234
     dumreg2 |  -.0316352   .3032999    -0.10   0.917    -.6312753     .568005
     dumreg3 |  -.2614979    .334916    -0.78   0.436    -.9236449     .400649
     dumreg4 |   -.336324   .2639313    -1.27   0.205    -.8581303    .1854822
       _cons |   15.11567   4.504893     3.36   0.001     6.209256    24.02209
------------------------------------------------------------------------------
```

. predict res, r

. gen res2=res*res

. gen le2=log(res2)

. gen one=1

Second Component of the Eq. (6.6)

Estimating the variance component of the stochastic production function (Second stage of Just and Pope Stochastic Production Function)

```
. reg le2 Sl np Prec Temp area one, noconstant

      Source |       SS       df       MS              Number of obs =     165
-------------+------------------------------           F(  6,   159) =   74.69
       Model |  4137.61334     6  376.146667           Prob > F      =  0.0000
    Residual |  775.532345   159  5.03592432           R-squared     =  0.8422
-------------+------------------------------           Adj R-squared =  0.8309
       Total |  4913.14569   165  29.7766405           Root MSE      =  2.2441

------------------------------------------------------------------------------
         le2 |      Coef.   Std. Err.      t    P>|t|     [95% Conf. Interval]
-------------+----------------------------------------------------------------
          Sl |  -1.370271   1.115674    -1.23   0.221    -3.574272    .8337295
          np |   7.859001   2.575332     3.05   0.003     2.771463    12.94654
        Prec |  -3.810539   2.942292    -1.30   0.197    -9.623001    2.001923
        Temp |  -12.12751   4.247575    -2.86   0.005    -20.51854   -3.736471
        area |   .6617005   .7510893     0.88   0.380    -.8220674    2.145468
         one |   21.75097   19.47758     1.12   0.266    -16.72675    60.22869
------------------------------------------------------------------------------
```

predict pderr
(option xb assumed; fitted values)

. predict errerr, r

*did not generate the Judge Matrix

. gen perr=exp(pderr)

. gen alpha=sqrt(perr)

. *now generate the weights for the regression

. gen int1=1

. gen tint1=int1/alpha

. gen td1=dumreg1/alpha

. gen td2=dumreg2/alpha

. gen td3=dumreg3/alpha

. gen td4=dumreg4/alpha

. gen tyld=yld/alpha

. gen tSl=Sl/alpha

. gen tnp=np/alpha

. gen tPrec=Prec/alpha

. gen tTemp=Temp/alpha

. gen tarea=area/alpha

. iis(risk)

Estimating the Third Stage of Just and Pope Stochastic Production Function

```
. xtgls tyld tint1 tSl tnp tPrec tTemp tarea td1 td2 td3 td4
Cross-sectional time-series FGLS regression

Coefficients:  generalized least squares
Panels:        homoskedastic
Correlation:   no autocorrelation

Estimated covariances      =         1          Number of obs      =       165
Estimated autocorrelations =         0          Number of groups   =        15
Estimated coefficients     =         6          Time periods       =        11
                                                Wald chi2(12)      =  81334.44
Log likelihood             = -345.3994          Prob > chi2        =    0.0000

------------------------------------------------------------------------------
       tlyld |      Coef.   Std. Err.      z    P>|z|     [95% Conf. Interval]
-------------+----------------------------------------------------------------
       tint1 |  -26.31921   16.12563    -1.63   0.103    -57.92487    5.286438
         tSl |   .2137016   .1053052     2.03   0.042     .0073072    .4200959
         tnp |     7.3251   3.532569     2.07   0.038     .4013908    14.24881
       tPrec |    .324097   .1607089     2.02   0.044     .0091133    .6390806
       tTemp |  -.0321212    .060054    -0.53   0.593    -.1498249    .0855825
       tarea |   5.034507   2.559929     1.97   0.049     .0171378    10.05188
       _cons |  -.6405359   1.337449    -0.48   0.632    -3.261888    1.980816
------------------------------------------------------------------------------
```

Or

```
. xtreg tyld tint1 tSl tnp tPrec tTemp tarea td1 td2 td3 td4, re
Cross-sectional time-series FGLS regression

Random-effects GLS regression                   Number of obs      =       165
Group variable (i): risk                        Number of groups   =        15

R-sq:  within  = 0.9969                         Obs per group: min =        11
       between = 0.9995                                        avg =      11.0
       overall = 0.9980                                        max =        11

Random effects u_i ~ Gaussian                   Wald chi2(12)      =  74926.27
corr(u_i, X)       = 0 (assumed)                Prob > chi2        =    0.0000

------------------------------------------------------------------------------
       tlyld |      Coef.   Std. Err.      z    P>|z|     [95% Conf. Interval]
-------------+----------------------------------------------------------------
       tint1 |  -26.31921   16.12563    -1.63   0.103    -57.92487    5.286438
         tSl |   .2137016   .1053052     2.03   0.042     .0073072    .4200959
         tnp |     7.3251   3.532569     2.07   0.038     .4013908    14.24881
       tPrec |    .324097   .1607089     2.02   0.044     .0091133    .6390806
       tTemp |  -.0321212    .060054    -0.53   0.593    -.1498249    .0855825
       tarea |   5.034507   2.559929     1.97   0.049     .0171378    10.05188
       _cons |  -.6405359   1.337449    -0.48   0.632    -3.261888    1.980816
-------------+----------------------------------------------------------------
     sigma_u |          0
     sigma_e |  1.9901802
         rho |          0   (fraction of variance due to u_i)
------------------------------------------------------------------------------
```

References

Adams RM, Hurd BH, Lenhart S, Leary N (1998) Effects of global climate change on agriculture: an interpretative review. Clim Res 11:19–30

Baltagi BH (2010) Econometric analysis of panel data, 4th edn. Wiley, West Sussex

Boubacar I (2010) The effects of drought on crop yields and yield variability in Sahel. In: The Southern Agricultural Economics Association annual meeting. The Southern Agricultural Economics Association, Orlando

Carew R, Smith EG, Grant C (2009) Factors influencing wheat yield and variability: evidence from Manitoba, Canada. J Agric Appl Econ 41(3):625–639

Chi-Chung C, McCarl BA, Schimmelpfennig DE (2004) Yield variability as influenced by climate: a statistical investigation. Clim Change 66:239–261

Cline WR (1996) The impact of global warming on agriculture: comment. Am Econ Rev 86(5):1309–1312

IPCC (2007) Climate change 2007: Synthesis report. Working group contribution to the Fourth assessment report of Intergovernmental Panel on Climate Change. Intergovernmental Panel on Climate Change (IPCC), Geneva

Isik M, Devadoss S (2006) An analysis of the impact of climate change on crop yields and yield variability. Appl Econ 38:835–844

Joshi NP, Maharjan KL, Piya L (2011) Effect of climate variables on yield of major food crops in Nepal: a time-series analysis. J Contemp India Stud Space Soc 1:19–26

Just RE, Pope RD (1979) Production function estimation and related risk considerations. Am J Agric Econ 61(2):276–284

Just RE, Pope RD (1978) Stochastic specification of production functions and economic implications. J Econ 7:67–86

Kim MK, Pang A (2009) Climate change impact on rice yield and production risk. J Rural Dev 32(2):17–29

Lobell DB, Asner GP (2003) Climate and management contributions to recent trends in U.S. agricultural yields. Science 299(5609):1032

Lobell DB, Field CB (2007) Global scale climate-crop yield relationships and the impacts of recent warming. Environ Res Lett 2(1):1–7

Lobell DB, Ortiz-Monasterio JI, Asner GP, Matson PA, Naylor RL, Falcon WP (2005) Analysis of wheat yield and climatic trends in Mexico. Field Crop Res 94:250–256

Mendelsohn R, Nordhaus WD, Shaw D (1994) The impact of global warming on agriculture: a Ricardian analysis. Am Econ Rev 84(4):753–771

Nicholls N (1997) Increased Australian wheat yield due to recent climate trends. Nature 387: 484–485

Peng S, Huang J, Sheehy JE, Laza RC, Visperas RM, Zhong X et al (2004) Rice yields decline with higher night temperature from global warming. Proc Natl Acad Sci 101(27):9971–9975

Porter JR, Semenov MA (2005) Crop responses to climatic variation. Philos Trans R Soc Biol Sci 360:2021–2035

Rosegrant MW, Ewing M, Yohe G, Ian Burton SH, Valmonte-Santos R (2008) Climate change and agriculture: threats and opportunities. German Technical Cooperation (GTZ), Eschborn

Saha A, Havenner A, Talpaz H (1997) Stochastic production function estimation: small sample properties of ML versus FGLS. Appl Econ 29:459–469

Schlenker W, Roberts MJ (2008) Estimating the impact of climate change on crop yields: the importance of non-linear temperature effects. NBER Working Paper 13799. National Bureau of Economic Research, Massachusetts

Takle ES, Pan Z (2005) Climate change and crop production: clallenges to modeling future scenarios. In: Lal R, Stewart BA, Uphoff N, Hansen DO (eds) Climate change and global food security. Taylor & Francis Group, Boca Raton, pp 383–403

Wang J, Mendelsohn R, Dinar A, Huang J, Rozelle S, Zhang L (2009) The impact of climate change on China's agriculture. Agric Econ 40:323–337

Chapter 7
Effects of Climate Change on Regional Agriculture Production, Food Price and Food Insecurity

Abstract Climate change is expected to have both negative as well as positive effect on agricultural production. Therefore, depending upon the regional characteristics of climate and its future direction some regions are going to be benefited by climate change in terms of agricultural productivity, whereas some regions' agriculture is exposed to a higher degree of vulnerability. Considering these differential regional effects of climate change on agriculture, this chapter discusses projected impact of climate change in agricultural productivity, price of agricultural commodities, and food insecurity in different regions.

Keywords Agriculture • Food insecurity • Price • Production • Productivity

7.1 Impact of Climate Change on Regional Agricultural Production

As discussed in the earlier section, agriculture is quite sensitive to small change in climatic factors, which differs significantly from region to region. Impact of change in climatic factors on agriculture also varies across the region over the world, some regions are benefiting from such changes while some regions are losing. Overall, the impact of climate change on global agricultural GDP will be between −1.5 % to +2.6 % by 2080, with considerable regional variation (Fischer et al. 2002b). For instance, high latitude regions (which specially cover developed countries) are expected to benefit from higher temperatures. Increase in temperature in such regions will expand the areas potentially suitable for cropping, the length of growing period will increase and crop yields may rise (Schmidhuber and Tubiello 2007).

Similarly, a moderate increase in temperature in some humid and temperate grassland my increases pasture productivity and reduce the need for housing and compound feed for livestock. In contrast to this, low-latitude regions (especially tropical developing countries) will be adversely impacted by the projected change

K.L. Maharjan and N.P. Joshi, *Climate Change, Agriculture and Rural Livelihoods in Developing Countries*, Advances in Asian Human-Environmental Research, DOI 10.1007/978-4-431-54343-5_7, in climate variables. It is expected that these regions will suffer increased heat waves and droughts. Therefore, some cultivated areas are expected to become unsuitable for cropping and some tropical grassland may become increasingly arid in such regions. The considerable increase (of around 160 million ha) in suitable cropland is estimated at higher latitudes (developed countries), whereas at lower latitudes (developing countries) there will decline in potential cropland by around 110 million ha. In addition, shifts in the quality of cropland in developing countries is expected. For instance, due to deterioration of land quality, land for double cropping would decline by between 10 and 20 million ha, and also land suitable for triple cropping would decline by 5–10 million ha in Sub-Saharan Africa alone (Fischer et al. 2002a). Such asymmetry would be further exacerbated due to the differences in adaptive capacity between developed and developing countries.

Besides, differences in the following factors will also affect agricultural production in different way across the region:

- The strength and saturation point of elevated CO_2 response of crops.
- Water quality, availability and irrigation.
- Crop interactions with air pollutants, weeds, pathogens and disease.
- Changes in the frequency in climate extremes and changes in mean climate.
- Timing and implementation of adaptation strategies, which will be dependent on the socioeconomic status of particular household, community, or region.

Acknowledging the fact there will be location specific impact of climate change on agriculture; there are several literatures that analyze regional impacts of climate change on agricultural production through yield of the crops. Virtually all of those analyses utilize the methodologies discussed in Chap. 6 to assess the climate change impact on the yield of agricultural crops. This section discusses the regional variations on impacts of climate change on agricultural production based on the literatures reviewed. Most of such studies however, are conducted in the developed countries case. There are very limited literatures on cases of developing countries.

Literatures based on the results of the crop simulation model and integrated assessments indicate that the impact of climate change on crop yield thereby food systems at the global scale will overall be small in the short run (first half of the twenty first century). However, the impact will turn more negative in the long run (after the first half of the twenty first century). It is mainly due to expected mean temperatures increase regionally and globally above 2.5 °C (Tubiello et al. 2008). Thus, there will be threshold effects. For instance, generally crop yield response to temperature increases of 2 °C rise is positive, whereas crop yield response 4 °C is negative (Rosenzweig et al. 1995) with more negative crop impacts in lower latitudes.

Cline (2007) illustrated that compared to the existing CO_2 concentration level; carbon fertilization would have the substantially more favorable results even when temperature increases (Figs. 7.1 and 7.2). The regional impact, however, follow the similar path. Both figures show that damages of temperature increase is concentrated mostly in the latitudes closer to the equator, whereas gains from such changes are mostly concentrated in latitudes closer to the poles. Most parts of the Southern Hemisphere will suffer nearly uniform losses except for New Zealand and certain

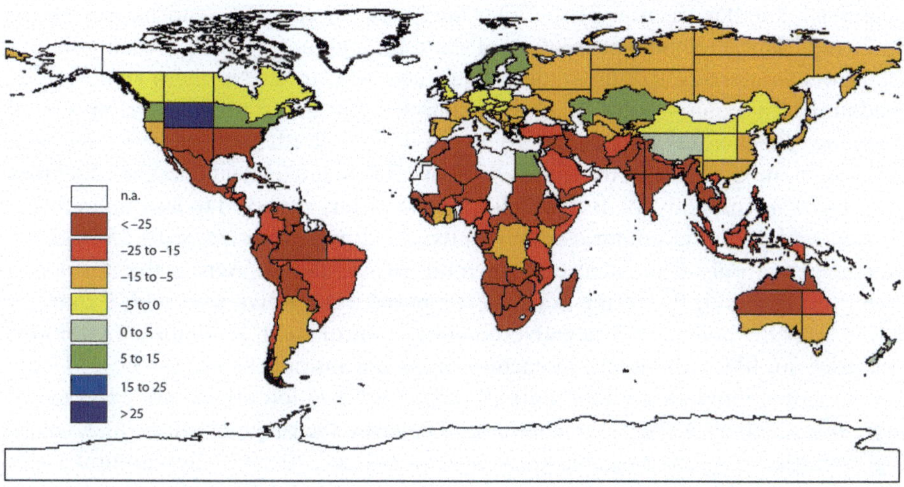

Fig. 7.1 Impact of temperature rise on agricultural productivity without carbon fertilization (in percentage) (Cline 2007). *Note*: n.a. refers to "not applicable" for Alaska, Northern Canada, and Antarctica and "not available" otherwise

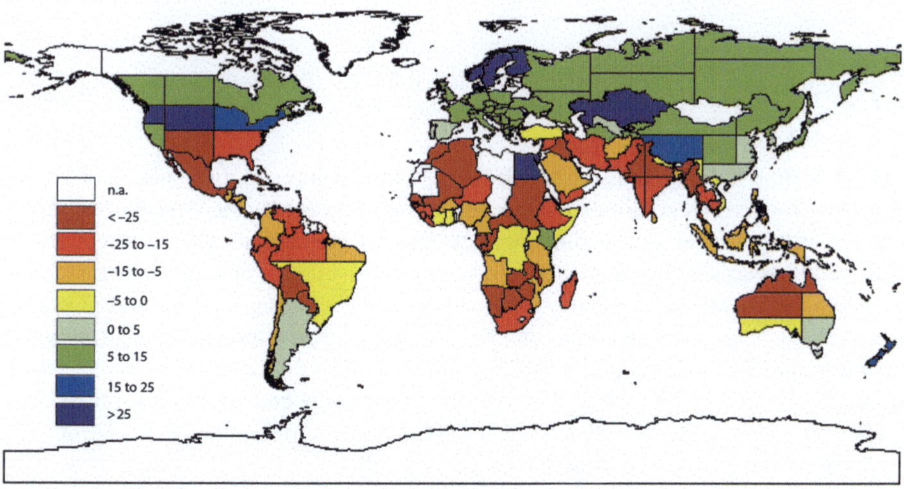

Fig. 7.2 Impact of temperature rise on agricultural productivity with carbon fertilization (in percentage) (Cline 2007). *Note*: n.a. refers to "not applicable" for Alaska, Northern Canada, and Antarctica and "not available" otherwise

parts of Argentina and Australia. This reflects the paucity of land masses in the latitudes above 35°S. This might cause a decline in agricultural productivity in developing countries, in the Southern Hemisphere, by between 9 % and 21 % (Cline 2007). However, the situation is just in reverse of the Northern Hemisphere. Land masses lying between the equator and about 35°N, where losses are predominant is of a much smaller fraction of the total land area. In contrast, land masses above 35°N represent a significant fraction of the total land area, which is also going to gain from an increase in temperature. Thus, countries in the Northern Hemisphere, predominantly composed of developed countries are going to benefit from global warming. Fischer et al. (2002a, b) showed that North America and Russian Federation, most of which territories lie above 35°N, are expected to gain 20–50 % and 40–70 %, respectively. But, sub-Saharan Africa is expected to lose up to 9 %.

Across developing countries, the results are largely negative, with 25 negative outcomes and only 6 positive outcomes out of 43 studies. Across the developed countries, however, the results lean more toward the positive side with 9 outright positive results and only 3 negative outcomes out of total 27 studies. Developed countries are likely to benefit in each of the scenarios considered (1 °C, 2 °C and 3.5 °C global temperature increase) as carbon fertilization effects are expected to more than compensate for climate effects. However, the range of effect on developing countries is between gains of 4 % to a loss of 20 % of agricultural GDP. Therefore, developing countries may be relatively worse off as climate change is expected to benefit developed countries. Thus, warming may result in slight reductions in agricultural prices as global aggregate supply expands. This will mildly harm farmers in developing countries, though they will have beneficial effects for consumers worldwide (Mendelsohn 2000).

7.1.1 Climate Change Impact in African Agriculture

Beside decline in land area suitable for agriculture crop production, yields of grains and other food crops could also decrease substantially across the African continent. Despite the potential production increase due to increase in CO_2 concentration, expected increases in the frequency of drought due to climate change is the main factor behind such decline in yield as well as land suitable for agriculture. There is also a high chance that some crops like maize, which is also an important food crop in the region, might be discontinued in some areas. Not only the crops, but also livestock will suffer from climate change due to deteriorated rangeland quality and changes in area from rangeland to unproductive shrub land and desert (Easterling et al. 2007).

Even within Africa, the impact might differ with the nature of farming as well as socio-economic status of the household. Taking the case of 11 African countries, Mendelsohn (2009), shows that a one degree centigrade increase in temperature would reduce average net revenue per hectare by 6 %. Further disintegration shows that the marginal temperature effect is −8 % in dryland farms, whereas it is positive (+3 %) in irrigated farms. This suggests that warming is harmful in rain-fed dryland farms but beneficial in irrigated farms. In the case of precipitation, 1 cm/month will increase farm net revenue by 7 %. Marginal effects of rainfall are positive in both dryland as well as irrigated farms. However, the increase is greater in rain-fed farms (+8 %) compared to irrigated farms (+3 %). Therefore, both kinds of farms are expected to benefit if the trend of rainfall increase in the region. But, decrease in rainfall as projected will adversely affect agriculture especially in dryland farms.

Mean yield responses as well as yield variability of maize, millet and sorghum in eight Sahelian countries show the harmful effects of increase in temperature, even though these crops are considered to be heat-tolerant. Rainfall, on the other hand, has a positive significant contribution to the increase of mean crop yields, i.e., increases in rainfall, evenly spread throughout the growing season, are beneficial to crops (Boubacar 2010). However, the persistent occurrence of droughts is hindering the yield growth of these crops, and is expected to reduce yield in future as well under the future climate change scenarios with increased incidence of drought.

7.1.2 Climate Change Impact in Latin American Agriculture

The length of the rainy season and the occurrence of extreme events (droughts and floods) have affected large areas in Latin America. Studies have revealed that in Latin America there is reductions in yield and increased variability in crop productivity (Adams et al. 1998). Except for wheat and sunflower in Argentina, yield of all other crops are projected to decline. All the crops (wheat, maize, and soybean) considered for Brazil, maize in Mexico, barley and wheat in Uruguay, and maize and soybean in Argentina will be adversely affected by climate change. The outcome of the GISS general circulation model for several locations in Latin America with predicted temperature increased of 3–4.5 °C and changes in rainfall of −10 % to 30 % suggests 10–30 % crop yield reduction in above four countries considered (Adams et al. 1998).

In Mexico, increases in minimum temperature as well as maximum temperature are having an adverse impact on wheat yield (Lobell et al. 2005). Any increase in these temperatures due to climate change, therefore, will reduce wheat yield in Mexico. However, in the case of other Latin American countries increases in summer temperature will increase the net revenue from farms, but the effects are clearly nonlinear. This means further temperature increase, in the long run will adversely affect agriculture in these countries (Mendelsohn 2009).

7.1.3 Climate Change Impact in Asian Agriculture

Asia is expected to suffer substantial decrease in cereal production potential by the end of this century as a consequence of climate change (Cruz et al. 2007). However, there will be significant regional differences in the yield responses of wheat, maize and rice to projected climate change. East and South-East Asia would benefit the yield increase up to 20 %, while Central and South Asia would suffer up to 30 % decline in the yield even if the direct positive physiological effects of CO_2 are taken into consideration. Crop yield is estimated to decrease by 2.5–10 % in the 2020s and 5–30 % in the 2050s in parts of Asia under the highest future emission trajectory situation (Parry et al. 2004). This could lead to 3.8 % decline in rice production by the end of the twenty first century in Asia (Murdiyarso 2000). The situation will be more severe in rainfed South and South-East Asia.

In South Asia, the increase in temperature beyond 2.5 °C will drop the yields of non-irrigated wheat and rice significantly incurring loss in farm-level net revenue between 9 % and 25 %. Thus, the net cereal production in South Asian countries is projected to decline at least between 4 % and 10 % by the end of this century, even under the most conservative climate change scenario (Lal 2007)

Country specific cases show that production of rice and wheat might drop by 8 % and 32 %, respectively in Bangladesh (Faisal and Parveen 2004). Similar is the case for India, where 0.5 °C rise in winter temperature would reduce wheat yield by 0.45 t per hectare (Lal et al. 1998). Also, it has been revealed that with the temperature rise of 0.5–1.5 °C, there will be 2–5 % decline in yield potential of wheat and maize in India (Aggarwal 2003). Ricardian analysis in Pakistan shows that the temperature increase in key growing seasons could be harmful resulting into annual crop farming losses ranging from US$100–200. Considering the fact that average crop net revenue in Pakistan is US$450, these impacts could be devastating for farmers (Ahmed and Schmitz 2011)

In China, as well, a mean air temperature increase of 2 °C could decrease rain-fed rice yield by 5–12 %. Similarly, 1 °C increase in wheat growing season temperature would reduce wheat yields by 3–10 % (You et al. 2009) However, impact of increased temperature in irrigated farms is expected to be positive. The net impact of climate change on Chinese agriculture will be only mildly harmful at first, but the damages will grow over time. The Northeast and Northwest region of China will have to bear the largest damages of climate change (Wang et al. 2009). In South Korea, though the temperature is positively related to average rice yield, it is also positively related with rice yield variability. Such increase in rice yield variability in Korea will cause fluctuation of rice production as well as rice price instability. It is estimated that climate change might result in an increase of rice yield variability by up to 10 % to 20 % (Kim and Pang 2009).

Examining India specifically, the Indian agronomic studies suggest that extensive warming could cause significant reductions in yields in the absence of adaptation and carbon fertilization. Grain yield would fall in India by 25–40 %, if temperatures rise by 4 °C. Rice yields would fall 15–25 % and wheat yields would fall 30–35 % for similar temperature increase (Mendelsohn 2000). The Ricardian results for India, which includes adaptation but not carbon fertilization, suggest only modest agricultural damage estimates. Although all the studies predict agricultural losses from warming, the cross-sectional studies find smaller losses than the agronomic studies. Using pooled analysis, Sanghi et al. study finds that a 2 °C warming would reduce average Indian net revenues by only about 4 %. Using the repeat annual analyses, Kumar and Parikh determine that a 2 °C warming would decrease revenues by about 8 %. Even with a 3.5 °C warming, the Sanghi et al. study find damages of only about 15 % while Kumar and Parikh predict damages of about 23 % (cited in Mendelsohn 2000). The different results from the agronomic-simulations and the cross-sectional studies could be due to the adaptation, as the private adaptation could reduce potential climate damages by between one-fourth and one-half. Besides, cross-sectional studies reveal that climate has important seasonal patterns in India. Net revenues in India decrease precipitously with a warmer winter, spring, and summer temperatures, whereas net revenues will increase with warmer fall temperature. The harmful effects of warmer spring and summer temperatures in India are expected given that temperatures are quite hot already in India during these periods. The effect of warmer fall in all locations is expected to be beneficial as the warmer temperatures help ripen and dry the harvest.

Net revenues are also sensitive to seasonal precipitation, but the effects are smaller and offsetting. Winter precipitations are beneficial but summer and springs are not. In India, additional summer rains are not helpful because most of India enjoys a monsoon during this period. The cross-sectional studies reveal that the effect of climate change is not uniform across India. Even if the warming was the same throughout the country, some areas would lose heavily, most would be moderately damaged, and some areas would even benefit slightly. Warming would most heavily damage the Western Coastal districts, whereas districts in several Eastern states along the coast would benefit.

It would appear that the more capital-intensive agricultural production systems are less sensitive to climate. The more capital-intensive systems appear to be able to substitute purchased inputs for climate more readily. Thus, developing countries are likely to be more sensitive to climate change than developed countries. Therefore, with the advancement in technology, the agriculture sectors in developing countries will become less sensitive over time. This means, adoption of new farm technologies may reduce some of the potential damages caused by climate change.

7.2 Impact of Climate Change on Agriculture Product Price

Change in yield, yield variability, and crop acreage of agricultural crops due to climate change will have a direct bearing on the supply of the product thereby its price. Since supply will have a direct negative relation to the price, any decline in supply will result in an increase in the price, *ceteris paribus*. Under the present context of a globalized world, prices of agricultural commodities are heavily influenced by changes in global food supplies. Therefore, any assessments of the effects of climate change on price of agricultural commodities in one country or region should reflect the changes in world supplies of those commodities. It is estimated that due to climate change the additional price increase would be 14.4 % (Tubiello et al. 2008). Consumers will be adversely affected by an increase in the price, however, such negative effects will partially or totally offset by producers' gains from higher prices. But in any cases, reduction in the total supply will results in a decline in total welfare. In the long run, higher prices stimulate producers to produce more and increase supply, which results in new equilibrium levels of prices and quantities.

Tubiello et al. (2008) lists three basic messages that emerge from studies on the likely impacts of climate change on food prices, which are as follows:

1. Overall, food prices are expected to rise moderately in line with moderate increases of temperature until 2050. There is also a change of a slight decline in real prices until 2050, but after 2050 with further increases in temperatures, prices are expected to increase more substantially.

2. For some commodities such as rice and sugar, prices are forecast to increase by as much as 80 % above their reference levels without climate change.
3. Expected changes in price due to increase in temperature are much smaller than the expected price changes from socioeconomic development paths. Under the SRES A2 scenario the price increase in real cereal price is about 170 %.

As the crop yields vary across the regions, the effects on price changes will also vary across the regions. For instance, developed countries especially the USA and Canada being the net exporter of agricultural crops, these countries are expected to be buoyed by both rising food as well as feed-grain prices (Adams et al. 1998). Whereas, developing countries, which imports cereal is estimated to be increased by 10–40 % by 2080, is the one to suffer from any increase in food price (Rosegrant et al. 2008).

7.3 Impact of Climate Change on Food Security

The Food and Agriculture Organization of the United Nations (FAO) defines food security as a situation that exists "when all people at all times have physical, and economic access to sufficient, safe and nutritious food that meets their dietary needs for an active and healthy life" (FAO 1996). Thus, food security has four dimensions namely; food availability, food accessibility, stability in availability and accessibility, and utilization. Climate change is expected to affect all of these four dimensions of food security.

As we discussed in an earlier section, climate change will affect food production (food availability) directly through changes in agro-ecological conditions. Such effect will be mixed and vary regionally. Due to the reduced production potential of tropical developing countries, which have poor land and water resources and are faced with serious food insecurity, the burden of food insecurity is expected to further increase in these countries. Globally, however, the potential for food production is projected to increase with an increase in average temperature between 1 °C and 3 °C, beyond this it is projected to decrease. However, increased urbanization and population in the developing world is likely to increase food demand that even surpasses the projected increase in agricultural production in the short run. An increase in yield variability with future expected trends of climate variables will hamper the stability in food supply (food availability). Moreover, changes in the patterns of extreme events, such as increased frequency and intensity of droughts or flooding, will affect the stability of food supply as well as accessibility due to the market response to such event resulting in increased food price.

Changes in food costs and the capacity to procure food are directly affected by changes in commodity supply and resultant price. Increase in price due to decline in supply will reduce consumption levels and adversely affect consumer welfare. The global cereal prices have been projected to increase more than three times by the 2080 because of decline in net productivity due to projected climate change (Parry et al. 2004). This will certainly lead to localized increase in food price, putting the subsistence farmer in greater risk as their accessibility for deficit food would be hampered by increased food price. Moreover, their main produce like sorghum, millet etc. is expected to suffer the dual risk of a potential drop in productivity as well as the danger of losing crop genetic diversity that has been preserved over generations. With the more pronounced and more widespread, droughts and floods there will be short-term fluctuations in food production in semiarid and sub-humid areas, more specifically in the sub-Saharan Africa and parts of South Asia. This means that poorest regions with the highest level of chronic undernourishment will also be exposed to the highest degree of instability in food production (food availability) (Schmidhuber and Tubiello 2007)

Utilization dimension of food security will be affected by climate change through the ability of individuals to utilize food effectively through an alteration in the conditions of food safety and by increasing the disease pressure from vectors, water and food-borne diseases (Schmidhuber and Tubiello 2007). Projected change in climate can initiate a vicious circle where infectious diseases cause or compound hunger through change in utilization of food consumed, which in turn makes the affected populations more susceptible to infectious diseases (Tubiello et al. 2008). For instance, increases in daily temperatures will raise the frequency of food poisoning, particularly in temperate regions. Similarly, populations in water-scarce regions are likely to face decreased water availability with implications of food processing and consumption, whereas in coastal areas there will be a higher risk of flooding of human settlements due to both sea level rise as well as increased heavy precipitation. Therefore, there might be a higher risk of increase in the number of people exposed to vector-borne (e.g. malaria and dengue) and water-borne (cholera) diseases, which lowers the capacity of the people to utilize food effectively (Easterling et al. 2007).

There are a number of studies based on complex modeling frameworks integrating the outputs of GCMs, agro-ecological zone data, dynamic crop models, and socioeconomic model in order to quantify the impacts of climate change on food security at regional and global scales. A number of limitations within these models make such projection highly uncertain. However, despite such limitations and uncertainties, there are several fairly robust findings emerged from these studies. For instance, climate change is expected to increase the number of people at risk of hunger compared to reference scenarios with no climate change (Easterling et al. 2007; Tubiello et al. 2008). The extent of risk will depend on the projected socioeconomic changes. It is estimated that climate change will increase the number of undernourished people in 2080 by between 5 and 10 million under SRES B1 scenario and 120–170 million people under SRES A1 scenario (Fischer et al. 2002a; Fischer et al. 2005) .

Schmidhuber and Tubiello (2007), summarizes that the climate change without CO_2 fertilization would reduce the number of undernourished people by 2080 by only around 20–140 million. The numbers vary based on different scenarios. For instance, under A2 scenario with no CO_2 fertilization, it is estimated that around 950–1300 million people will remain undernourished in 2080, whereas the number could be 740–850 million with CO_2 effect on crops. The number might increase under the growing competition between food production and bio-energy in the coming decades and centuries.

At present, Asia represents the region as the most food-insecure region. The risk of hunger is likely to remain high in the future as well. It is estimated that there will be additional 49 million people at risk of hunger by 2020 under A2 scenario without carbon fertilization. The figure may go up to 132 million in 2050 and 266 million in 2080 (Parry et al. 2004). Africa, more specifically sub-Saharan Africa, could replace Asia and become the most food-insecure region by 2080. This is largely independent of climate change and is mostly the result of the projected socioeconomic development (path) in the region (Easterling et al. 2007). For the entire SRES and climate change scenarios, it is expected that sub-Saharan Africa might account 40–50 % of global hunger by 2080, compared to around 24 % at present. Even some analyses show that sub-Saharan Africa might count up to 70–75 % of the world's total food-insecure by 2080. Regional variations in food insecurity could be better explained by population changes than impact on food availability (Parry et al. 2005). Therefore, economic and other development policies will be critical in influencing the impact of climate change on food security.

The chapters hereafter (Chaps. 8–10) will focus on specific cases of Nepal based on our own research works. Chapter 8 deals with the particular issue of climate change in Nepal. The chapter discusses GHG emission trends, temperature and precipitation trends, impacts of climate change on different sectors and its relation to poverty, and prospects of mitigation practices in revenue generations. Chapter 9, on the other hand, assesses the impact of climate change on yields of major food crops based on the national data on temperature, precipitation, and yield data using the time series data from 1978 to 2008. Finally, Chap. 10 analyzes the public opinion and perception on climate change based on household survey data. Such perceptions are then triangulated with observed data in the locality.

References

Adams RM, Hurd BH, Lenhart S, Leary N (1998) Effects of global climate change on agriculture: an interpretative review. Clim Res 11:19–30

Aggarwal PK (2003) Impact of climate change on Indian agriculture. J Plant Biol 30:189–198

Ahmed MN, Schmitz PM (2011) Using Ricardian technique to estimate the impacts of climate change on crop farming in Pakistan. In: The EAAE 2011 congress, Zurich, 30 August to 2 September 2011

Boubacar I (2010) The effects of drought on crop yields and yield variability in Sahel. In: The Southern Agricultural Economics Association annual meeting. The Southern Agricultural Economics Association, Orlando

Cline WR (2007) Global warming and agriculture: impact estimates by country. Peterson Institute for International Economics, Washington, DC

Cruz RV, Harasawa H, Lal M, Wu S (2007) Asia. In: Parry ML, Canziani OF, Palutikof JP, van der Linden PJ, Hansen CE (eds) Climate change 2007: impacts, adaptation and vulnerability. Contribution of Working Group II to the Fourth Assessment Report of the Intergovernmental Panel on Climate Change. Cambridge University Press, Cambridge, pp 469–506

Easterling WE, Aggarwal PK, Batima P, Brander KM, Erda L, Howden SM, Tubiello FN (2007) Food, fibre and forest products. In: Parry ML, Canziani OF, Palutikof PJJP, Hanson CE (eds) Climate change 2007: impacts, adaptation and vulnerability. Contribution of Working Group II to the Fourth Assessment Report of the Intergovernmental Panel on Climate Change. Cambridge University Press, Cambridge, pp 273–313

Faisal IM, Parveen S (2004) Food security in the face of climate change, population growth and resource constraints; implications for Bangladesh. Environ Manage 34:487–498

FAO (1996) Rome declaration on world food security and world food summit plan of action. World Food Summit 13–17 November, 1996. Food and Agriculture Organization of the United Nations, Rome

Fischer G, Shah M, van Velthuizen H (2002a) Climate change and agricultural vulnerability. A special report prepared as a contribution to the World Summit on Sustainable Development. International Institute for Applied System Analysis, Laxenburg

Fischer G, Shah M, Tubiello FN, van Velthuizen H (2005) Socio-economic and climate change impacts on agriculture: an integrated assessment, 1990–2080. Philos Trans R Soc Biol Sci 360:2067–2083

Fischer G, van Velthuizen H, Shah M, Nachtergaele FO (2002b) Global agro-ecological assessment for agriculture in the 21st century: methodology and results. Research Report RR-02-02. International Institute for Applied System Analysis, Laxenburg

Kim MK, Pang A (2009) Climate change impact on rice yield and production risk. J Rural Dev 32(2):17–29

Lal M (2007) Implications of climate change on agricultural productivity and food security in South Asia. Key vulnerable regions and climate change—identifying thresholds for impacts and adaptation in relation to Article 2 of the UNFCCC. Springer, Dordrecht

Lal M, Singh KK, Rathore LS, Srinivasan G, Saseendran SA (1998) Vulnerability of rice and wheat yields in NW—India to future changes in climate. Agr Forest Meteorol 89:101–114

Lobell DB, Ortiz-Monasterio JI, Asner GP, Matson PA, Naylor RL, Falcon WP (2005) Analysis of wheat yield and climatic trends in Mexico. Field Crop Res 94:250–256

Mendelsohn R (2000) Measuring the effect of climate change on developing country agriculture. In: Two essays on climate change and agriculture: A developing country perspective (FAO Economic and Social Development Paper 145). Food and Agriculture Organization of the United Nations, Rome, pp 2–31

Mendelsohn R (2009) The impact of climate change on agriculture in developing countries. J Nat Resour Policy Res 1(1):5–19

Murdiyarso D (2000) Adaptation to climatic variability and change: Asian perspectives on agriculture and food security. Environ Monit Assess 61:123–131

Parry ML, Rosenzweig C, Iglesias A, Livermore M, Fischer G (2004) Effects of climate change on global food production under SRES emissions and socio-economic scenarios. Glob Environ Change 14:53–67

Parry M, Rosenzweig C, Livermore M (2005) Climate change, global food supply and risk of hunger. Philos Trans R Soc Biol Sci 360:2125–2138

Rosegrant MW, Ewing M, Yohe G, Ian Burton SH, Valmonte-Santos R (2008) Climate change and agriculture: threats and opportunities. German Technical Cooperation (GTZ), Eschborn

Rosenzweig C, Parry ML, Fischer G (1995) World food supply. In: Strzepek KM, Smith JB (eds) As climate changes: international impacts and implications. Cambridge University Press, Cambridge, pp 27–56

Schmidhuber J, Tubiello FN (2007) Global food security under climate change. Proc Natl Acad Sci 104(50):19703–19708

Tubiello F, Schmidhuber J, Howden M, Neofotis PG, Park S, Fernandes E, Thapa D (2008) Climate change response strategies for agriculture: challenges and opportunities for the 21st century. Agriculture and Rural Development Discussion Paper. The World Bank, Washington, DC

Wang J, Mendelsohn R, Dinar A, Huang J, Rozelle S, Zhang L (2009) The impact of climate change on China's agriculture. Agr Econ 40:323–337

You L, Rosegrant MW, Wood S, Sun D (2009) Impact of growing season temperature on wheat productivity in China. Agr Forest Meteorol 149:1009–1014

Chapter 8
Climate Change in Nepalese Context: Impacts, Mitigation Issues, and Relation with Poverty

Abstract This chapter discusses the different dimensions of climate change in Nepal. Nepal has a negligible share of global Green House Gases (GHGs) emission but increasing at significantly higher rate compared to its fast growing neighboring economies like China, India and Bangladesh. Sector-wise emission shows that agriculture and forestry are two most important sectors contributing almost 90 % of the total emissions. Hence, any mitigation effort in Nepal should consider these two sectors, which are also the most important sectors for poor people. Consequently, intervention in these sectors will help to build rural community's resilience to Climate Change (CC). Increase in temperature and variable rainfall pattern have a negative direct influence on water resources at the highest level followed by agriculture, forest, and health sectors of the country. Being signatories of major international legislations related to CC, Nepal has a prospect to generate revenue through mitigation effort, which could be used to deal with adverse impact caused by CC. Alternative energy promotion, forest management, and agricultural practice are potential areas, which can generate revenue from carbon trading. All these prospective areas have multiple functions of mitigation, adaptation as well as economic empowerment of the vulnerable section of the population.

Keywords Agriculture • Alternative energy • Forestry • Vector-borne diseases • Water resources

---

This chapter draws from PhD dissertation of Niraj Prakash JOSHI (2011). Findings discussed in this chapter was presented in the International Conference on Climate Change, Livelihoods, and Food Security, June 9–10, 2009, Institute of Development Studies, Jaipur, India in collaboration with Association of Asian Scholars, Bangkok. Part of this chapter is also published in Journal of International Development and Cooperation, 18(4), Special Issue: 21–35.

K.L. Maharjan and N.P. Joshi, *Climate Change, Agriculture and Rural Livelihoods in Developing Countries*, Advances in Asian Human-Environmental Research,
DOI 10.1007/978-4-431-54343-5_8, © Springer Japan 2013

8.1 Introduction

Clear indication of CC in the earth has been reported. Over the last few decades, the temperature of the earth's surface has been rising and predicted to rise further if the proper attention is not paid. This has caused changes in weather patterns, melting of glaciers and rise in sea level. In addition, more frequent storm events, increased events of drought, increased number of El-Nino and other adverse climatic situations can also be attributed to the global CC. Prediction shows that rise in 2 °C temperature is inevitable even if emissions are reduced to less than 50 % of the current level by 2050. This increase in temperature is determined to be "an upper limit beyond which the risks of grave damage to ecosystems, and of non-linear responses, are expected to increase rapidly". However, the current trend of emission i.e., emission well above 2000 levels in 2100, would lead to a 4 °C increase in temperature causing unavoidable devastating losses, and excessively higher adaptation costs (IPCC 2007c).

Such adverse CC put all countries in the vulnerable situation through increased stress in the economy as a whole. But the poorest countries and the poorest people within them are the most vulnerable as they are dependent on natural resources to a greater extent. In this decade alone around 3.5 billion people, almost all from developing and least developed countries, are likely to be affected by climate related disasters. This figure is significantly higher compared to approximately 0.8 billion in 1970s, 1.4 billion in 1980s and 1.9 billion in 1990s. During 1990s, around 200 million people per year were affected by climate related disasters in developing countries, in contrast to around a million in developed countries. Based on this, the World Bank estimates that people in developing countries are affected at 20 times higher the rate of those in developed countries (WB 2007a, 2008).

Degree of vulnerability to Nepal is even higher due to its rugged, steep topography, and fragile geological conditions, which make the country disaster prone. Besides, marginal population with low income, limited institutional capacity and greater reliance on climate-sensitive sectors like agriculture make the country subjected to higher risk (Regmi and Adhikari 2007; WB 2008). Therefore, exposure to risks and low adaptive capacity to cope with those risks are major factors contributing to vulnerable situation of the country to CC. This justifies the strong need of understanding local CC, and related hazards in order to develop mitigation and adaptation programs to minimize risks at different levels. Adaptation measure is urgently needed to reduce the impacts of CC particularly for the most vulnerable section of the population; therefore adaptation aspects should not be overlooked. However, as adaptations for CC require huge economic resources and is at an early stage of development, mitigation is highly cost effective and relevant in the long term together with possible revenue generation for the developing countries like Nepal (Dhakal 2001; IPCC 2007a). Therefore, this chapter aims to analyze different aspects of CC in Nepal namely; emission scenarios, CC scenarios, impacts of CC on poverty, and initiatives taken by Nepal and their prospects to generate revenues from international CC regimes in relation to their possible impact on poverty reduction.

8.2 Greenhouse Gas Emission Situation in Nepal

Carbon dioxide (CO_2), methane (CH_4), and nitrous oxide (N_2O) are the major GHGs that contribute to CC in Nepal and, therefore included in national GHG inventory of 1990/1991 and 1994/1995 (Dhakal 2001; MoPE and UNEP 2004). Nepal has a very negligible share (0.025 %) of global GHG emission (MoE 2011). However, if we see the trend of emission, it is increasing at a higher rate in Nepal. There was a 63.5 % increase in CO_2 emission between the first inventory period (1990/1991) and the second inventory period (1994/1995), which shows annual growth rates of 13.1 % (Fig. 8.1). Similarly, annual growth rate of 9.3 % per capita CO_2 emission was reported between 1990 and 2004 in Nepal, which is significantly high compared to its economically fast growing neighboring countries like India (3 %), China (4.4 %), and Bangladesh (4.4 %) (UNDP 2007). Growing consumption of fossil fuel is the main reason for such increase. Fuel consumption increased from 4,000 barrels per day to 7,258 barrels per day during 1990–1995 and reached 17,200 barrels per day in 2007, which indicates the annual growth rate of 16 % between 1990/1991 and 1994/1995 (EIA 2009).

Mainly, rice production, livestock, and biomass burning are responsible for CH_4 emission in Nepal (Fig. 8.1). Methane emission shows some positive signs in terms of emission reduction. This is mainly due to significant reduction in emission from rice production, and biomass burning or manure management. In addition, the promotion of minimum tillage farming in rice cultivation such as System of Rice Intensification (SRI) and visible reduction in rice production area from 1.4 million ha in 1990/1991 to 1.3 million ha in 1994/1995 could have contributed to the

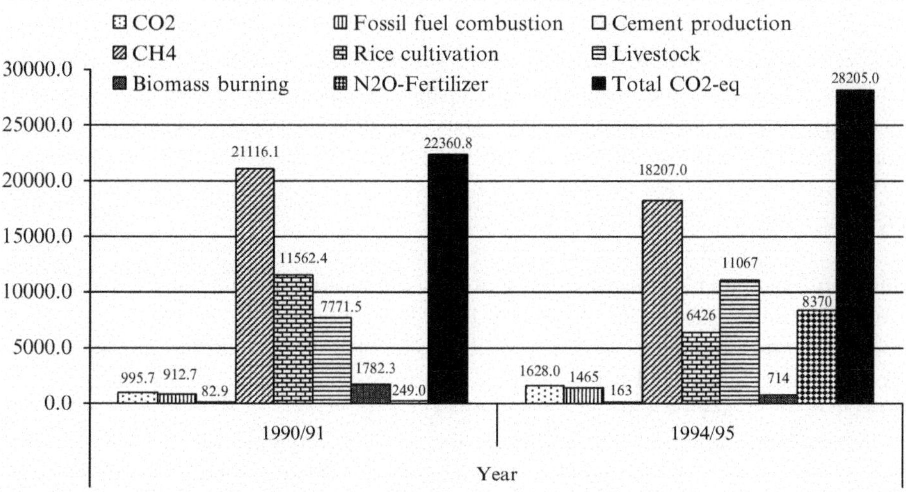

Fig. 8.1 GHGs emission from different sectors in Nepal from 1990/1991 to 1994/1995 (CO_2-eq in '000 tons') (Dhakal 2001; MoPE and UNEP 2004; IPCC 2007b)

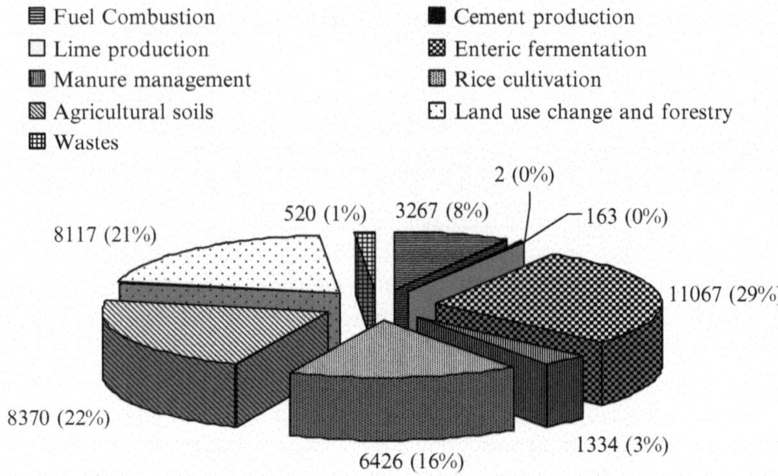

Fig. 8.2 CO_2-eq emission from different sectors in Nepal 1994/95 (in '000 tons') (MoPE and UNEP 2004; IPCC 2007b)

reduction of CH_4 emission. However, a rising population of ruminant livestock in the same periods resulted in an increase in CH_4 emission from livestock through enteric fermentation (MoAC 2005). At the same time manure management as well as replacement of fuel wood achieved through installation of 11,941 biogas plants between 1992/1993 and 1994/1995 could be the reason behind the significant reduction in CH_4 emission from biomass burning (Laudari 2008).

The significant increase in consumption of nitrogen fertilizer has resulted in a drastic increase in the emission of N_2O, despite reduction in the area under rice cultivation. Annual sales of urea increased from 81,000 to 121,000 tons between 1990/1991 and 1994/1995 (MoAC 2005). Rice cultivation in Nepal is mostly done under a submerged condition. Therefore, the increased use of nitrogen fertilizer (urea) triggered N_2O emission between these periods. However, thereafter there is a continuous decline in the use of nitrogen fertilizer and reached only around 7,000 tons in 2003/2004 (MoAC 2005). This may signify that there is a decline in N_2O emission since 1994/1995 in Nepal.

The rise in total CO_2-eq emission in Nepal can be observed between the same duration. The total CO_2-eq emission reached 28.2 million tons CO_2-eq in 1994/1995 from 22.4 million tons CO_2-eq in 1990/91. This indicates the increase in CO_2-eq emission at the annual rate of 5.8 %. With the inclusion of important GHGs sources like land use change and forestry, wastes and lime production, GHG emissions in 1994/1995 reached 39.3 million tons of CO_2-eq.

Agriculture has significant bearing on the total CO_2-eq emission. Enteric fermentation in livestock (29 %), manure management (3 %), rice cultivation (16 %), and agricultural soils (22 %) as components of agriculture altogether emit around 69 % of total CO_2-eq emission. This is followed by land use change and forestry, which contribute around 21 % of the total CO_2-eq emission, fuel combustion (8 %) and waste (1 %) (Fig. 8.2). This suggests the importance of agriculture and forestry sectors in any mitigation effort to reduce GHGs emission as well as building resilience to CC among the farmers. Since agriculture and forestry are the most important sources of livelihood for the majority of the poor in the country, mitigation measures on these sectors will have a high significance in reducing emission as well as poverty. Further, adaptation measure on these sectors in the short term is very crucial to deal with vulnerability caused by CC.

8.3 Indication of Climate Change

Indication of CC can be assessed mainly in terms of variations in temperature, and precipitation (Shrestha et al. 2000; IPCC 2007c). Five regional headquarters were chosen considering the altitude, coverage of geographical area, and consistency of data availability to analyze the changes in temperature and rainfall. These locations represent the country from the east to far-west, and altitudinal variation of 720 m above sea level (masl) in Surkhet to 2,310 masl in Dipayal. Other locations include Dhankuta (1,445 masl), Kathmandu (1,336 masl), and Pokhara (827 masl). Data on temperature is available from 1976 to 2005 for all locations except Dipayal for which data is available only from 1982 to 2005. Temperature variability is assessed in terms of annual as well as seasonal trends, whereas precipitation variability is assessed in terms of annual trend only in these five regions. Winter temperature is calculated based on average temperature for the month of December (of the preceding year), January and February. Similarly, summer temperature is calculated based on average temperature for the month of June, July and August.

Nepal has experienced the fastest long-term increase in temperature with 1.6 °C increase between 1976 and 2005 (Fig. 8.3f), which is very high compared to global temperature increase of 0.6 °C in the last three decades (IPCC 2007c). Trend analysis shows that temperature is increasing at an annual rate of 0.054 °C, which is statistically significant (Table 8.1). The rate is higher in winter (0.06 °C) compared to summer (0.05 °C). Moreover, several climate models in Nepal show that the warming trend will continue throughout the twenty first century (Table 8.2).

The highest rate of increase in annual temperature is recorded in Dhankuta, followed by Kathmandu, and Dipayal all of which has a relatively high altitude. Surkhet and Pokhara have relatively lower trend coefficients but still significant. This, in some extent, supports temperature increase faster at higher altitudes than at lower altitudes (Agrawala et al. 2003). Similar scenario can be observed from the seasonal breakdown of regional temperature trends as well. Dhankuta has the highest coefficient for both winter and summer temperature trends followed by Kathmandu and Dipayal. In all the cases we found that temperature increase in winter is higher compared to that of summer. The similar trend will continue in the coming days as well (Table 8.2). Therefore, people are now experiencing hotter summers and warm winters. Similarly, Nepal is experiencing increasing warm days and nights (Baidya and Karmacharya 2007). Figure 8.3 shows the detail temperature trends for all five locations including the national average. In all these cases we can see that temperature is in increasing trend in all locations.

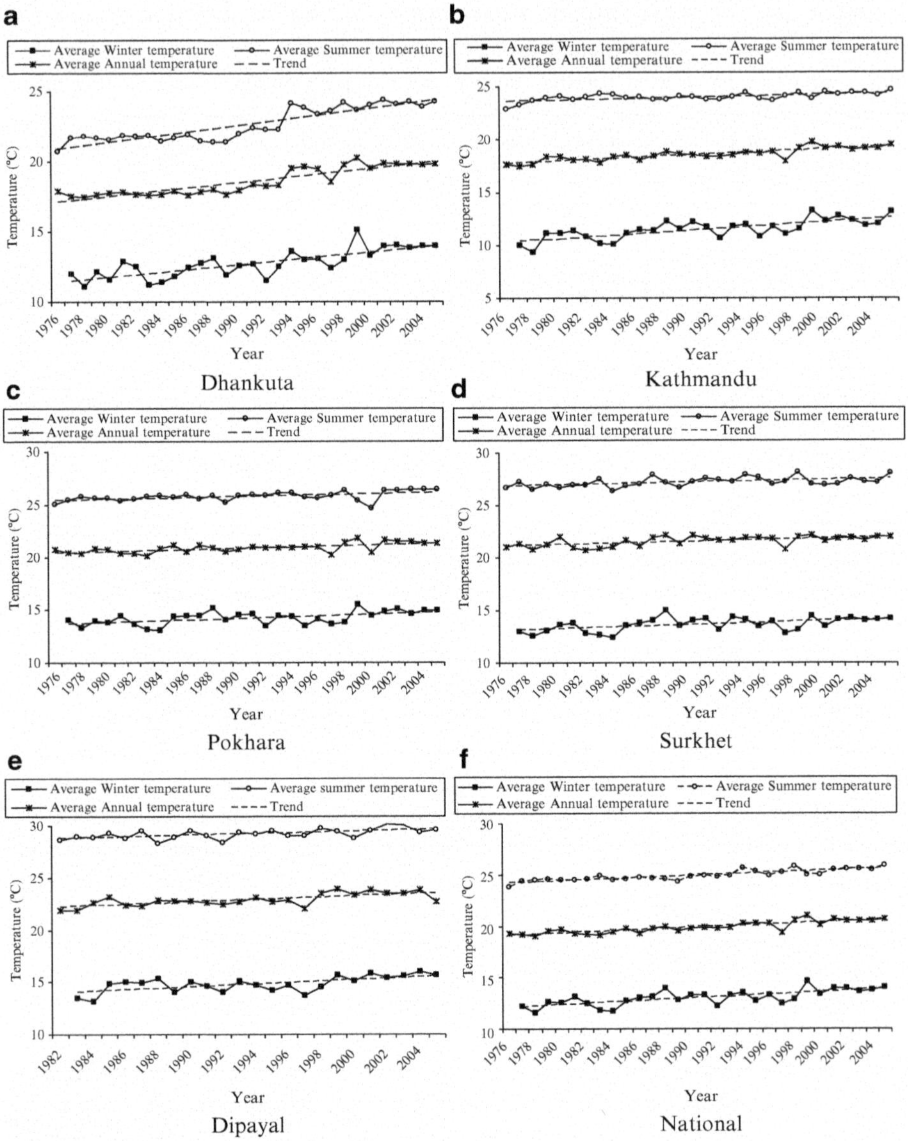

Fig. 8.3 Temperature trends in Nepal

Similar to the temperature, overall global precipitation has also increased by about 2 % since the beginning of the twentieth century, which is statistically significant. However, such increase is neither spatially nor temporally uniform. Indian monsoonal rainfall shows the increasing trend since 1974 (IPCC 2001). Since Indian monsoonal rainfall is the main source of precipitation in Nepal, it also experienced an increasing trend of precipitation, though it is very erratic over the years

Table 8.1 Coefficient of temperature and rainfall trend in Nepal by region (CBS 1987, 1993, 1997, 2005, 2007)

Variables	Coefficient	R^2 value	*P*-value
National			
Annual temperature	0.054	0.74	0.00***
Winter temperature	0.06	0.50	0.00***
Summer temperature	0.05	0.70	0.00***
Rainfall	6.1	0.1	0.1*
Dhankuta			
Annual temperature	0.1	0.79	0.00***
Winter temperature	0.12	0.60	0.00***
Summer temperature	0.09	0.80	0.00***
Rainfall	1.08	0.01	0.79
Kathmandu			
Annual temperature	0.06	0.71	0.00***
Winter temperature	0.08	0.54	0.00***
Summer temperature	0.03	0.38	0.01***
Rainfall	6.0	0.07	0.1*
Pokhara			
Annual temperature	0.03	0.41	0.00***
Winter temperature	0.04	0.28	0.00***
Summer temperature	0.02	0.26	0.00***
Rainfall	13.14	0.05	0.21
Surkhet			
Annual temperature	0.03	0.3	0.00***
Winter temperature	0.04	0.25	0.01***
Summer temperature	0.02	0.23	0.01***
Rainfall	4.1	0.02	0.49
Dipayal			
Annual temperature	0.06	0.46	0.00***
Winter temperature	0.07	0.38	0.00***
Summer temperature	0.04	0.36	0.00***
Rainfall	−8.3	0.06	0.26

Note: *** and * significant at 1 % and 10 % level of significance, respectively

Table 8.2 Prediction of temperature and precipitation in Nepal (GCM estimates) (Agrawala et al. 2003)

	Mean temperature increase (°C)			Mean precipitation increase (mm)		
Year	Annual	Winter	Summer	Annual	Winter	Summer
Baseline average	–	–	–	1,433	73	894
2030	1.2 (0.27)	1.3 (0.4)	1.1 (0.2)	71.6 (3.8)	0.6 (9.9)	81.4 (7.1)
2050	1.7 (0.39)	1.8 (0.58)	1.6 (0.29)	104.6 (5.6)	0.9 (14.4)	117.1 (10.3)
2100	3.0 (0.67)	3.2 (1.00)	2.9 (0.51)	180.6 (9.7)	1.5 (25.0)	204.7 (17.9)

Note: Figures in parentheses indicate standard deviation

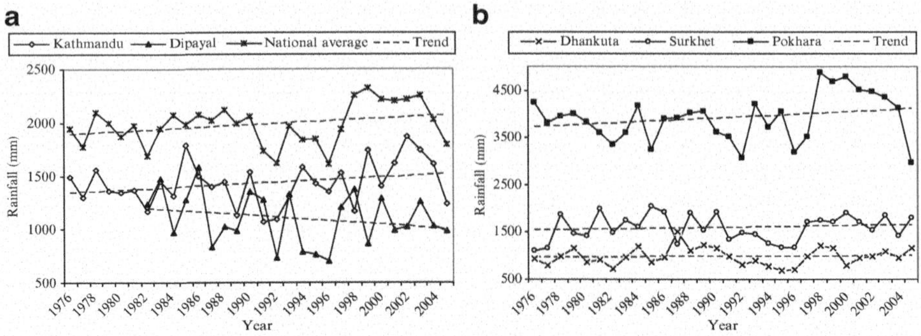

Fig. 8.4 Annual rainfalls in Nepal by different location

(Fig. 8.4). A R^2 value is very low for all cases of rainfall trend analyses. Coefficient of trend value shows that annual national average rainfall is increasing. The increase is statistically significant at a 10 % level of significance (Table 8.1). Such an increase can be attributed to the global warming, which results in an increase in land-ocean thermal contrast, thereby intensifying monsoon circulation (Shrestha et al. 2000). In addition, General Circulation Model (GCM) estimates an overall increase of precipitation in Nepal (Table 8.2). Seasonal breakdown of estimated precipitation shows that monsoon rain is going to be more intense, whereas dry season will be drier. This will be further intensified under a CO_2 doubling condition. In recent days, increasing heavy rainfall event as well as a maximum 24-h rainfall in the country is in increasing trend (Baidya and Karmacharya 2007). Monsoon rain, which contributes around 80 % of total rainfall, is the main source of waterborne disaster in Nepal. Therefore, increase in intensity of summer monsoon can be translated into an increase in intensity of water borne disasters like flood, landslide, and sedimentation. Consequently, there will be an enormous loss of settlements, infrastructures, and fertile top soil that lowers agricultural productivity. These are regular phenomena but becoming more intense in recent years.

In addition, the topography of a location also dictates rainfall patterns in Nepal. Trend analysis shows the decreasing trend of rainfall in Dipayal from Far-Western Hills of the country, but the coefficient is statistically non-significant. Dipayal is a location having the highest altitude. However, rainfall is continuously increasing in Kathmandu at a significant rate. Similarly, rainfall is continuously increasing in Pokhara, Surkhet, and Dhankuta but the increase is not statistically significant. Pokhara has the highest coefficient, and due to its typical topography it also receives the highest annual rainfall in Nepal.

8.4 Impact of Climate Change on Poverty

Poverty is persistent and widespread in Nepal, more specifically in the rural areas. Poverty measurements collected since 1977 show no indication of poverty reduction in the country. The sign of improvement in poverty reduction was realized only in the NLSS II (2003/2004) and NLSS III (2010/2011) (Joshi et al. 2010; CBS 2011). Despite such decrease in poverty, the nature of poverty however, still remains the same. Poverty is more rampant, deeper and severe in rural areas, and much worse in Mid-Western and Far-Western Hills and Mountains. Similarly, most of the poorest of the poor belong to the *Dalit* (oppressed caste), and ethnic communities. Agriculture wage laboring, casual laboring and self-employments in agriculture are the main sources of livelihoods of the poor in rural areas. Thus, poverty in Nepal is complex, diverse in nature, and associated with location, gender, caste/ethnicity, land ownership, occupation and low economic growth of the country. Consequently, due to persistent poverty in the country, there is a lack of institutional capacity to adapt with any adverse impact of climate change in Nepal despite being prone to natural disaster due to its rugged terrain with steep topography and fragile geological condition.

Nepal has a very negligible contribution on the global CC as it has a negligible share on global GHG emission and also has one of the lowest per capita GHGs emissions in the world (Olivier and Peters 2010; MoE 2011). But it is not free from adverse impact of CC. Its fragile geography, predominantly natural resource based livelihoods, and low level of adaptive capacity due to higher incidence of poverty place the country among the fourth most vulnerable country to CC (Maplecroft 2010).

Water resource, agriculture, forestry and biodiversity, and human health are some of the important sectors, which could be adversely affected by CC and consequently aggravate poverty in Nepal. Figure 8.5 shows how the two aspects of CC, namely; temperature and precipitation along with their extreme events, affect poverty in Nepal through water resource, agriculture, forestry and biodiversity, and health sector. Water resource, which is the most important resource of Nepal having the highest economic potential in terms of hydroelectricity generation as well as irrigation management, is ranked as the most vulnerable sector to CC in Nepal (Agrawala et al. 2003). This sector is affected mainly through variability in temperatures and precipitation. Rise in temperature in Nepal has an adverse impact on 3,252 glaciers covering a total area of 5,323 square kilometers (km^2) (ICIMOD and UNEP 2001). These glaciers are retreating at a faster rate compared to any other glaciers, and the rate is even higher compared to previous estimates (Pokhrel 2007). For instance, The Rika Samba Glacier in the Dhaulagiri region is retreating at a rate of 10 m per year. There are 2,323 glacial lakes in Nepal that cover an area of 75.7 km^2, of which 20 are reported to be dangerously close to bursting because of global warming (ICIMOD and UNEP 2001). It is calculated that up to 70 % of snow and glacier in the glaciated area above 5,000 m may disappear with the temperature increase of 4 °C (MoPE and UNEP 2004). Disappearance of glacier and snow consequently leads to the development of more glacial lakes or swelling of existing glacial lakes and increase potential GLOF hazards in Nepal.

A GLOF comes with enormous destruction. It poses threats to downstream settlements, infrastructure, natural resource, and human lives. Nepal has already experienced 25 GLOFs in the past (Gum et al. 2009). The Dig Tsho GLOF that occurred in 1985 was the most devastating one. It caused a 10–15 m high surge of water and debris to flood down the Bhote Koshi, and Dudh Koshi River for 90 km in Eastern Nepal. The flood swept the newly built Namche Small Hydel Project, 14 bridges,

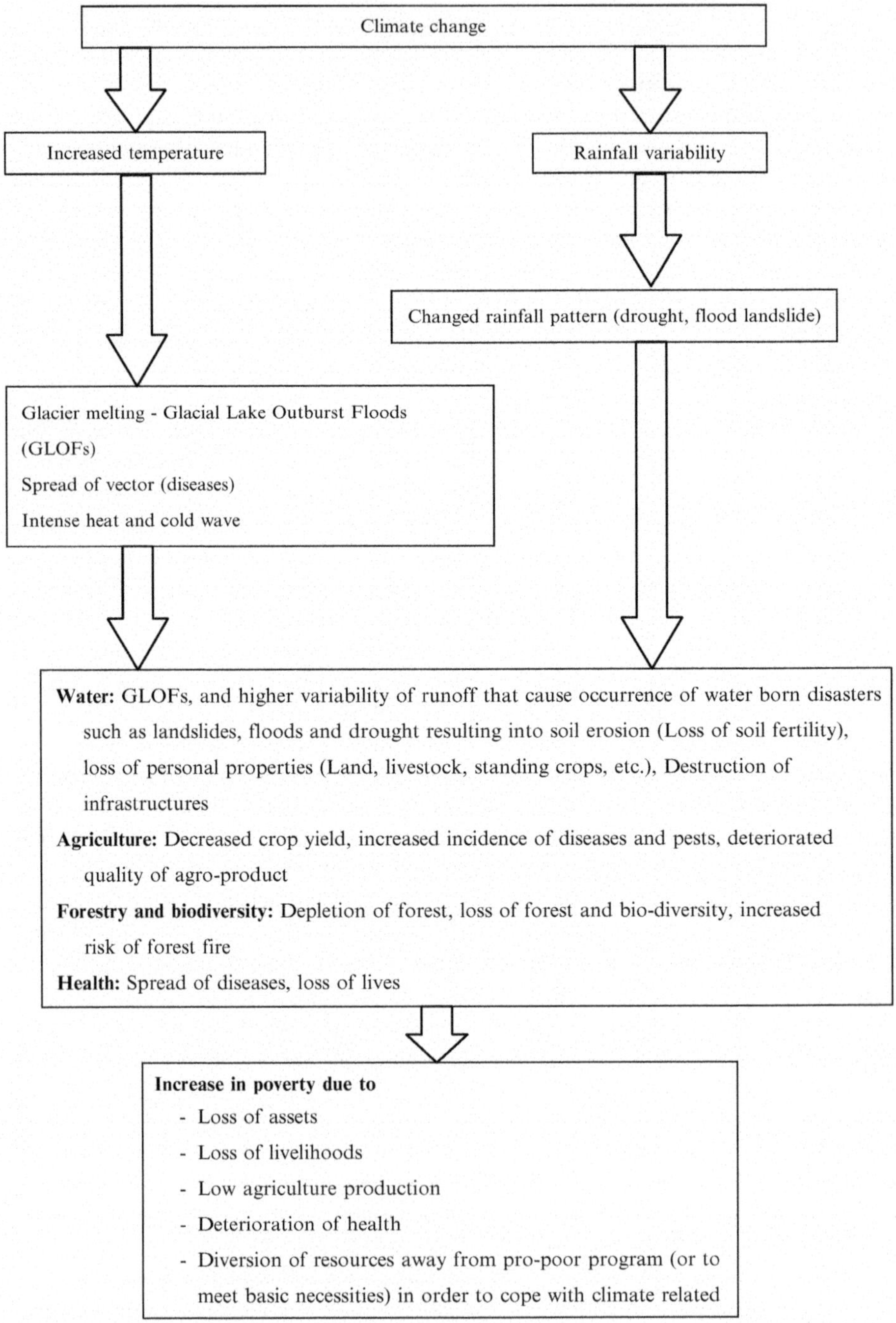

Fig. 8.5 Effect of climate change on poverty (Joshi 2011)

wide areas of cultivated land, at least 30 houses among others including livestock and inhabitants (Rana et al. 2000; Alam and Regmi 2004; Regmi and Adhikari 2007). Just recently, a collaborative anticipatory planning and management by the government, donors, and experts in GLOF mitigation is able to reduce the risk of a GLOF from the Tsho Rolpa Glacial Lake. This is the biggest glacial lake situated in the Rolwaling Valley of Eastern Nepal covering an area 1.76 km^2. Unless the mitigation effort was taken, the glacial lake would have caused significant destruction claiming more than 10,000 human lives, and huge infrastructure loss including 60 MW Khimti Hydropower (Rana et al. 2000).

The higher variability of runoff is another important factor in Nepal that can lead to increased water disaster such as flood, landslide and sedimentation, and more pronounced variations in water availability throughout the year. The available surface water of Nepal is 202 Cubic Kilometers (km^3), which goes down to only 26 km^3 in dry season (MoE 2010). The uneven distribution of rainfall and glacier retreat is the main reason for such variation that leads to water borne disaster. More than 80 % rainfall occurs between June and September through monsoon rain that comes from the Bay of Bengal. The current trend shows that the monsoon period is shortening, but at the same time the amount of rainfall is increasing, which means monsoon rain is becoming more intense. This is causing the problem of flood and landslide in the wet season and severe drought in the dry season. The widespread impact of change in hydrological flows has been observed in Nepal. It has impacted many irrigation systems, water-powered grain mills, hydropower plants and drinking water supply systems throughout the country (Gum et al. 2009). People are experiencing more intensive rainfall and subsequent flood and landslide that have a direct adverse impact on livelihood assets such as physical, natural, financial, social, and human especially among the poor (Vidal 2006; Gautam et al. 2007a, b; Pokhrel 2007). Therefore, water resource has high significance on the overall livelihood of the majority through a number of ways including disasters, hydropower that supplies around 91 % of the nation's power, irrigation, transportation and several other infrastructures.

Agriculture is another important sector to be hard hit by CC as it can be linked to the impact of CC on water, forests, health, and soil temperature. Considering its importance in Nepalese economy, any adverse impact on agriculture will jeopardize the life of many people. Around 66 % of the population (MoAC 2006), for whom agriculture is the mainstay, will face the risk of food insecurity due to CC. Since agriculture is heavily dependent on weather condition, this sector will be adversely affected through extreme rainfall, which results in increased runoff variability, soil fertility loss, temperature rise, as well as drought.

Nepalese agriculture is predominantly rain-fed. Therefore, any variations in rainfall patterns will have a direct impact on its agriculture. For instance, drought condition will result in decreased crop yields thereby total crop production. In 2005, food production of the country was adversely affected by drought that has caused 2 % and 3.3 % decrease in paddy and wheat production, respectively. Nearly, 10 % of agricultural land was left fallow due to rain deficit. Similarly, in 2006, drought in Eastern Tarai resulted in a decrease in rice production between 27 % and 39 %

(Regmi 2007). There were 21 % decline in rice production and 3 % decline in millet production in the same year. This dragged the country under food self-insufficiency for the first time since it started attaining food self-sufficiency in 1999 (Joshi et al. 2010). Moreover, World Food Programme (2010) identified climate related natural disasters like drought, flood, hailstorm, late/early rain, landslide, and crop pest & disease as the major causes of high or severe levels of food insecurity in a number of districts in the Far-Western Hills. These natural disasters have caused crop losses at 30–70 % among over 50 % of households (WFP 2010).

Drought became even worse in 2008/2009, which is considered one of the worst in the country's history with least rainfall and widespread across Nepal. It has resulted in decline in production of major winter crops; wheat and barley by 14.5 % and 17.3 %, respectively. The situation was even worse in some districts of Mid- and Far-West region. They received less than 50 % of average rainfall during the period of November 2008 to February 2009 (WFP 2010). Consequently, crop yields dropped by more than half. Thereby, many farmers are exposed to high risk of food insecurity as agriculture still remains subsistence in nature. In contrast, excessive rainfall also results in more frequent flood events that not only inundate the agriculture field and destroy the crops, but also destroy farmland and irrigation facilities. This consequently results in decreased agriculture production. Heavy rain and subsequent floods, landslides, and soil erosion are regular phenomena in Mid-Western Tarai, and Western regions of the country. Increased variability in runoff, therefore, is the major source of soil erosion in Nepal that washes away the fertile top soil in the sloppy areas, and sedimentation in inundated land. In both cases, soil fertility loss is the major outcome that consequently leads to production loss in agriculture, which also indicates loss of livelihood for the people who predominantly depends on agriculture.

Agriculture, being part of life science, will also respond to changes in temperature. Rise in temperature will affects agriculture through an increase in incidence of pests and diseases, and decrease in physiological performance of animal and poultry, thereby reducing crop and animal productions (IPCC 2007d). However, degree of effects will vary depending upon the altitudes. It is reported that the rise in temperature under atmospheric CO_2 doubling will initially increase the yield of rice, wheat, and maize in all three ecological regions of Nepal; Mountains, Hills, and Tarai. However, the rise in temperature at 4 °C will cause loss in rice and wheat yield in Tarai, which is considered the grain basket of the country having the highest proportion of land area under cultivation. Although the continued increase in yield is reported in Hills and Mountains, it will be obtained at the cost of exhausted soil fertility and likely adverse impacts on the nutritional value of crops. Increase in temperature under increased availability of atmospheric CO_2 leads to a vigorous growth of food-crops and reduce the level of soil organic carbon, soil micronutrient, and enhances decomposition by activating the microbial population in the soil thereby decreasing agricultural productivity in the long run (Malla 2003). Similarly, temperature rise by 2 °C would decrease the quality of meat and milk, hatchability of poultry, and increases the possibility of disease in the livestock (IPCC 2007d). Besides, temperature rise above 4 °C is detrimental to the existence of life on earth.

Joshi et al. (2011a) has shown that at the national level the current trends of climate variables had suppressed the yield growth of major food crops in Nepal. Suppression of yield is more pronounced in summer food crops like maize and potato. At the regional level, Joshi et al. (2011b) has shown that the adverse impact on yield of major food crops is more prevalent in low lying Tarai.

Forestry and health are the other sectors to be adversely affected by CC. Changes in temperature and precipitation would alter vegetation patterns of forests. It may cause a forest modification through migration of plant and animal species along with other biotic species towards the Polar Regions, changes in their composition, extinction of species, etc. With the increase in temperature, shifting upward of several domestic and wild plants and animal species has been reported in Nepal (Malla 2008). A study has shown that out of the 15 types of forest categorized by Holdridge model existing in Nepal under current CO_2 condition; three types will disappear if CO_2 concentration is doubled. Tropical wet forest and warm temperate rain forest would disappear, and cool temperate vegetation would turn into warm temperate vegetation (MoPE and UNEP 2004). Such change in vegetation would affect biodiversity in forests of Nepal. In addition, landslides, floods, and water erosions have resulted in massive depletion of forest. At the same time, summer drying and drought increased the risk of forest fire that poses threat to adjacent human settlements. Forest being an integral part of livelihood, such depletion of forest as well as loss of biodiversity will hamper the livelihood of the majority of the total population who are dependent on forest based livelihoods especially ethnic forest dwellers in rural Nepal like Chepangs (Piya et al. 2011).

Climate change has been recognized as one of the major challenges by the World Health Organization (WHO) for the health policy makers, planners, and managers and urged to address the issues before it becomes too late IPCC (2007d) projects an increase in under-nutrition and related disorders, morbidity and mortality due to heat waves, floods, droughts, windstorms, and fire. Similarly, the incidence of vector-borne diseases such as malaria, kalaazar, Japanese encephalitis, and dengue in tropical and sub-tropical regions, diarrheal diseases, and cardio-vascular diseases due to increase in ground-level ozone is expected to increase with the higher intensity of CC. In the particular case of Nepal, the vector-borne diseases are now moving to new regions as mosquitos from Tarai and Mid-Hills are being able to survive in the High-Hills as well. In 2006, 7 out of 13 Mountain districts of Nepal were classified as malaria prone districts due to the spread of the vector in these areas (WHO, n.d.). Similarly, incidence of kalaazar is now reported in more than dozens of Tarai districts. In addition, people in Nepal are exposed to death threats due to heat and cold waves. These extreme temperatures claimed more than 60 lives in 2003, which then rose to more than 110 in 2004 (Fig. 8.6). Since then, these extreme temperatures are continuously claiming lives. Given that less than one fifth of the population has access to modern health services, vulnerability to future CC in the health sector is quite high.

Therefore, the loss of shelter and infrastructure (physical assets), spread of vector-borne diseases and loss of lives (human assets), displacement of community (social assets), loss of water sources and cultivable land (natural assets), lower

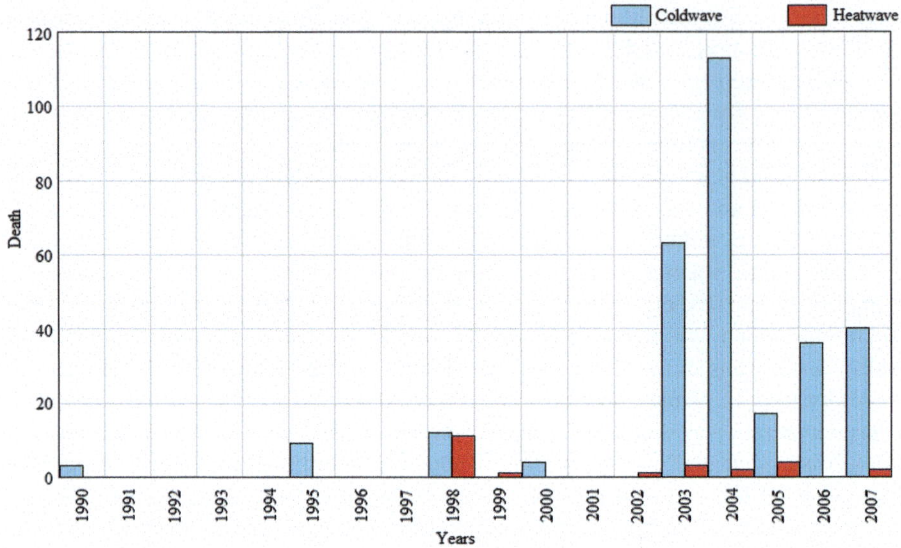

Fig. 8.6 Deaths caused by cold and heat waves in Nepal (http://online.desinventar.org/desinventar/index.php?r=NPL-1250695185-nepal_historic_inventory_of_disasters)

saving and higher debt (financial assets) are widespread evidences of CC impact in Nepal. All of these factors are responsible for higher vulnerability to CC, which exacerbates the problem of poverty especially among the marginal populations, who have very limited resources to cope with the problem.

8.5 Opportunities Created by International Climate Change Regimes for Poverty Reduction

Nepal is signatory of the United Nations Framework Convention on Climate Change (UNFCCC) and the Kyoto Protocol (KP), which are very crucial international legislations with regards to CC. However, the government of Nepal has not yet internalized aspects of CC in its policy documents. It was only in the Tenth Periodic Plan (2002–2007), that the government of Nepal committed itself to implement treaties on CC and took initiative to assess and control hazards caused by GLOFs taking CC convention as the basis (NPC 2003). The plan also envisaged the poverty reduction by optimal use of natural resources through community participation. The Eleventh Periodic Plan (2007–2010) moves forward in this direction and identify the promotion of carbon trade to achieve benefit from Clean Development Mechanism (CDM) under the KP. The promotion of alternative energy, and management of natural resources especially forest have been identified as a means for generating financial resources in the long term through carbon trading (NPC 2007). Similarly, possibilities of revenue generation from international CC regimes have been raised by Climate Change Policy, 2011 through low carbon and climate resilient development path (MoE 2011).

Household biogas, micro/mini hydropower, solar energy, Improved Water Mill (IWM) and Improved Cooking Stove (ICS) are some of the prospective alternative energy projects identified for international carbon trading. Two biogas projects that cover 19,396 biogas plants have already been registered in Clean Development Mechanism-Executive Board (CDM-EB) for carbon trading under voluntary basis in December 27, 2005, and have started generating revenue. With the estimated net emission reduction of 4.99 tons CO_2-eq/biogas-plant/year, and given US$7/ton CO_2-eq of carbon price (Koch-Mathian 2010), these projects are generating annual income of approximately US$0.65 million until 2012, the end of the first commitment period of the KP. Such revenue generated is expected to reduce dependency on large subsidies provided by the government and external donors. Also, such revenue will help to expand biogas installation in more remote and poorer areas of Nepal.

Considering the importance of biogas in tackling poverty, such expansion of biogas in remote and poorer areas of Nepal can also help in reducing poverty to some extent. Biogas plant can help to alleviate poverty through time saving (approximately 4 h/day) and cash saving. Time can be saved from shortening the time for cooking as well as saving time involved in collecting fuel-wood which would have been used for longer cooking periods. Similarly, cash savings of NRs. 25,499/HH/year can be achieved through the replacement of kerosene for lighting, reduced use of fertilizer, reduced expenses in health due to better sanitation, significantly lower indoor air pollution compared to fuel-wood used for cooking and kerosene used for lighting. Until 2006/07, there are 185,585 biogas plant installed in the country of which 96.2 % are operational. Therefore, any effort to bring 178,533 operating biogas plants under the small scale CDM project will generate around US$6.2 million per year. This amount can be utilized for scaling up biogas installation among the poor through subsidy and credit. Thus, it will further support to achieve poverty reduction goal of the country. Moreover, the existing number of biogas installation is only 10 % of total potential (Laudari 2008). Therefore, there is a high scope for dissemination of the installation all over the country especially in rural areas.

The Nepal Micro-Hydro Project (MHP) is the second CDM project in Nepal to be registered under CDM-EB, and the Energy Reduction Purchase Agreement (ERPA)was signed on June 29, 2007 by Alternative Energy Promotion Center. Thus, there is a possibility to generate Certified Emission Reduction (CER) of 324,999 tons CO_2-eq through the promotion of 15 MW MHPs by the end of the project year 2012. Out of these, 191,000 tons CO_2-eq could be sold at the rate of US$10.25/ton of CO_2-eq (WB 2007b). The price difference for CER from biogas and MHP is mainly due to the risk associated with the project as well as demand and supply situation from the particular project (Ascui and Costa 2007; Castillo 2007). Thus, MHPs being more mature and reliable compared to the biogas, the price offered for MHP is higher. MHPs will be developed under Rural Energy Development Programme (REDP); therefore, will operate in the poorest and geographically isolated areas serving the marginalized groups in rural Nepal through provision of off-grid electricity. This will provide a large number of rural households with electricity for lighting, milling and other needs. Thus, the project will help in poverty alleviation through employment generation as well as direct local environmental benefits through reduction in diesel and kerosene consumption (reduced CO_2 emission), and the use of dry cells (lowering chemical pollution and health hazard) and lead acid cell batteries (reducing pollution and transport cost involved in charging) (WB 2007b). Similarly, Project Idea Note (PIN) for IWM has already been submitted to the Designated National Authority (DNA), and that for ICS was supposed to be submitted to DNA on July 2008. These two CDM projects are very crucial from a viewpoint of poverty reduction as they have rural orientation. Besides, the PIN is being prepared for solar energy projects, electric vehicle, landfill solid waste management, and the vertical shaft brick kiln.

Forestry is another important sector in which about two thirds of the globe's terrestrial carbon is sequestered in the form of standing forest, forest understory plant, leaf and forest debris, and in forest soils together with other non-natural stocks. Nepal has 39.6 % of the total area covered by forest. Under the forestry sector, the CDM mechanism of the KP recognizes the afforestation and reforestation project to be eligible for carbon trading for the first commitment period. However, despite substantial plantation activities through community-based forest management and leasehold forest management programs, forestry-based CDM has not been initiated in Nepal so far. Therefore, some of the Community Forestry (CF) projects, leasehold forestry projects, private lands, and national forest have potentials to be brought under the CDM mechanism if such projects are developed to meet the necessary criteria. They should meet at least the following three criteria: plantation area equal to or greater than 0.5 ha, the crown coverage should be less than 10 %, and the plantation carried out in 2000 onwards in areas where there had been no forests since 1990.

Most of the forest regeneration activities are taking place in the hilly regions of the country. Therefore, carbon sequestration studies done by the International Centre for Integrated Mountain Development (ICIMOD) in 2007 in the Himalayas including Nepal could be very much relevant to the Nepalese context in general. The carbon sequestration capacity of Nepalese forest is 6.89 tons CO_2/ha/year (Banskota et al. 2007). It is estimated that Nepal can negotiate the price of at least US$5/ton CO_2 for the carbon sequestration by Nepalese forest. Food and Agriculture Organization (2006) reported that in Nepal plantation activities were carried out in 52,000 and 53,000 ha of land on 2000 and 2005, respectively. This also means that Nepal can claim US$3.6 million from the plantation activities in degraded land if such plantation was carried out with due consideration to bring under the CDM mechanism. Plantation activities in Nepal basically took place on private land, community forest, leasehold forest, and government forest. Similarly, by July 2000, plantation on 8,000 ha was done through leasehold forestry, which has reached to 17,244 ha in 2007 (FAO 2000; DoF 2007). This indicates that between 2000 and 2007, some 9,244 ha of degraded land was brought under plantation through the leasehold forestry program. Thus, the modest calculation shows that Nepal can generate revenue of around US$0.32 million per year only through leasehold forest, which could go up significantly if thorough study is made in this direction. Considering the success of the program in tackling poverty through secure right of land and employment generation (FAO 2006; DoF 2007), such revenue could be crucial in scaling up the program in around 1.6 million ha of barren lands or grasslands with scattered trees. Thus, afforestation and reforestation project eligible for CDM have economic potential of around US$55 million together with its contribution on an overarching goal of poverty reduction from its extension in all potential areas.

Exclusion of projects on natural forest conservation under the category of "avoided deforestation" hinders the possibility of bringing CF of Nepal under the CDM mechanism. However, the recognition of avoiding deforestation by the international community for its higher carbon mitigation benefits and sustainability has raised the prospects of Nepalese CF for carbon trading (IPCC 2007c). Considering a wide coverage of CF and protected areas in Nepal, it could be an important sector for revenue generation through carbon trading. In addition, given that deforestation is being the single most important source of carbon emission, there is an unequivocal emphasis to curb deforestation in developing countries as part of future responses to CC. The international community made an agreement in this direction during the 13th Conference of Parties (COP13) of the UNFCCC in Bali in 2007. The proposed Reducing Emissions from Deforestation and forest Degradation (REDD) policy is a new international legal framework for CC mitigation, which was emerged during the COP13, and at the present it is undergoing vigorous discussions. As it recognizes forest as carbon sources, management of existing forests, and rights of indigenous people who are dependent on forest resources to meet their subsistence needs, it is appealing for carbon trading. It is also considered as a "road map" for post-Kyoto Protocol after 2012 on the role of forests in the global climate budget. It has a provision of compensating developing countries in proportion to the amount of carbon emission that are reduced by halting its national deforestation rate below the baseline.

The World Bank launched a forest carbon fund for the REDD initiative called "Forest Carbon Partnership Fund" (FCPF). This fund has the dual objectives of building capacity for REDD in developing countries, and testing a program of performance-based incentive payments in some pilot countries. The FCPF can also be regarded as a precursor to the REDD (Karky and Banskota 2009). Nepal is one of the 13 tropical countries whose Readiness-Project Idea Note (R-PIN) is selected under this fund. After the formulation of the full Readiness Plan and its approval, Nepal will be able to implement a prototype of REDD and gain experience and build capacity to operationalize REDD by taking on board CF in an experimental way under FCPF. In addition, Nepal has successfully started generating revenue from CF of three watersheds in Dolkha, Gorkha and Chitwan with the implementation of the first-ever pilot Forest Carbon Trust Fund. A total sum of US$95000 was handed over to representatives from those three watersheds as the payments for their successful effort to sequestrate additional 0.1 million tons of CO_2 in 2011 compared to that of 2010 from around 10,000 ha of CF (ICIMOD 2011). This shows that any successful initiative of Nepal to implement REDD would fetch as much as US$42.7 million from CF and US$82.4 million from protected areas (2.4 million ha excluding buffer zones as these areas are also covered by CF to some extent) annually. Besides, there are also several hectares of land under private ownership, which could be brought under either CDM or REDD.

Agriculture in Nepal is predominantly subsistence in nature, with the very low level of external input use such as fertilizers, irrigation, pesticides and improved seeds. Due to heavy energy required for these external inputs especially fertilizer and irrigation, any effort to reduce the use of these resources or efficient use of these resources could significantly reduce GHG emissions from agriculture. Moreover, development and promotion of agricultural system that built on local resources for production input will be crucial in building resilience to CC. In this line, SRI with the baseline of methane emission, and Organic Agriculture (OA) with the baseline of nitrous oxide emission could be prospective projects to be brought under CDM. Steps to bring SRI under CDM are already in progress. Therefore, Nepal can claim its share from more than 1,000 ha of area under SRI, which is expanding at higher rate in Nepal from Tarai (60 masl) to Mid-Hills (around 2,000 masl) (Upreti 2008). Besides, considering tolerance of SRI to adverse climatic influences such as drought, storms, hot spells and cold snaps that results in reduced economic and agronomic risk, and higher yield compared to conventional practice, SRI could be an important practice of adaptation to CC (Uphoff 2007). Organic agriculture also serves in this direction. Love Green Nepal, an NGO, has taken an initiative to incorporate OA in the carbon market (UNEP, n.d.). However, there is a lack of documentation for area under OA as well as their certification mechanism. Any initiative to bring these practices under CDM through research and development not only generates the carbon credit but also helps to adapt against the adverse impacts of CC among the resource poor farmers.

8.6 Conclusion

Climate change is an unequivocal fact the earth is already experiencing, caused mainly due to the anthropological GHGs emission. CC is regarded as the greatest threats posed to the humankind putting more pressure on the poorest countries and the poorest people therein. Nepal has rugged terrain with steep topography and fragile geological conditions as well as higher incidence of poverty. The country is prone to disaster amidst limited institutional capacity and greater reliance on climate-sensitive sectors like agriculture thereby quite sensitive to any adverse change in climate. Therefore, this paper dealt with different aspects of CC and its relation with poverty. Nepal has a negligible share of global GHGs emission. However, the rate of emission increase is high mainly due to the constantly increasing use of fossil fuels, emission from livestock, and use of N_2O fertilizer. Sector-wise emission shows that agriculture and forestry are two most important sectors contributing almost 90 % of the total emissions in Nepal. Therefore, any mitigation efforts in Nepal should consider these two sectors, which are also the most important sectors for poor people to build resilience to CC. Despite such low level of emission, Nepal has already shown some indication of CC in terms of rising temperature, variability in rainfall and more frequent occurrence of climate related natural disasters.

Increase in temperature and rainfall patterns have a negative direct influence on water resources at the highest level followed by agriculture, forestry, and health sectors of the country. Increased risk of GLOFs poses threat to important infrastructures and settlements downstream claiming properties and lives of thousands. Similarly, higher variability in run-off results in increased water disasters such as flood, landslide, sedimentation, and variations in water availability throughout the year. These kinds of events adversely affect livelihood assets. Agriculture is also adversely affected by variations in temperature and rainfall. Crop loss due to flooding, inundation, landslide, and drought is a common phenomenon in Nepal these days reducing the production of major crops. The impact of CC in the forest is an alteration of forest composition and thereby loss in biodiversity. Increased temperature will cause migration of forest species towards the Polar Regions thereby loss of three forest types. Also, Nepal is experiencing depletion of forest land due to landslides, floods, water erosion, and forest fires. The spread of vector-borne disease to the new regions is the major challenge in the health sector as a consequence of CC. In addition, increased incidence of waterborne disaster and negative impact on agriculture and forest will affect the health sector adversely.

Being a signatory of major international legislations related to CC, Nepal has a prospect to generate revenue through mitigation effort, which could be used to deal with adverse impact caused by CC. Alternative energy promotion, forest management, and sustainable agricultural practice are potential areas, which can generate revenue from carbon trading. Two biogas projects and one micro-hydro project have been successfully registered and have started generating revenue through CDCF/World bank. Biogas projects for CDM cover only 10 % of installed biogas plants in the country, which itself is only 10 % of the potential biogas installations. Therefore, there is huge potential to generate revenue from the biogas sector through CDM mechanism. Similar is the case from micro-hydro projects as the country is rich in water resources. In the forest sector, leasehold forest and part of community forest has the prospect to generate revenue from the CDM mechanism under afforestation and reforestation provisions. In the context that Nepal is in the process to prepare full readiness plan that enables it to implement a prototype of REDD, any successful initiative to implement REDD policy would generate more than US$122 million from the forest sector alone. SRI and OA are two sustainable agricultural practices that have the potential to be brought under the CDM mechanism. These agricultural practices can also be regarded as adaptive measures against adverse impacts of CC as they reduce dependency on external inputs as well as improve tolerance against adverse weather conditions. All these prospective areas have multiple functions of mitigation, adaptation as well as economic empowerment of the vulnerable section of the population. Therefore, a proactive role of Nepal in international forum with research and development to incorporate these aspects in international negotiations and capacity development of its own in the field is very crucial to deal with adverse impacts of CC and meet its overarching goal of poverty reduction as well. In addition, further research on the impact of climate variables on agriculture based on the historical evidence as well as livelihood of rural poor based on the household level data is highly recommended.

References

Agrawala S, Raksakulthai V, van Aalst M, Larsen P, Smith J, Reynolds J (2003) Development and climate change in Nepal: focus on water resources and hydropower. Organization for Economic Co-operation and Development, Paris

Alam MA, Regmi BR (2004) Adverse impacts of climate change on development of Nepal: integrating adaptation into policies. Capacity Strengthening in the Least Developed Countries for Adaptation to Climate Change, Bangladesh Center for Advanced Studies, Bangladesh

Ascui F, Costa PM (2007) CER pricing and risk. In: Hodes G, Kamel K (eds) Equal exchange: determining a fair price for carbon. UNEP Risø Centre, Denmark, pp 78–93

Baidya SK, Karmacharya J (2007) Observed and future climate change in Nepal. In: National workshop on climate change and human health: potential impact, vulnerability and adaptation in Nepal. Nepal Health Research Council, Kathmandu

Banskota K, Karky BS, Skutsch M (2007) Reducing carbon emissions through community managed forests in the Himalayas. International Centre for Integrated Mountain Development, Kathmandu

Castillo MP (2007) Market perspectives to determine for fair carbon price. In: Hodes G, Kamel S (eds) Equal exchange: determining a fair price for carbon. UNEP Risø Centre, Denmark, pp 94–99

CBS (1987) Statistical year book of Nepal, 1987. Central Bureau of Statistics (CBS), Kathmandu

CBS (1993) Statistical year book of Nepal, 1993. Central Bureau of Statistics (CBS), Kathmandu

CBS (1997) Statistical year book of Nepal, 1997. Central Bureau of Statistics (CBS), Kathmandu

CBS (2005) Statistical year book of Nepal, 2005. Central Bureau of Statistics (CBS), Kathmandu

CBS (2007) Statistical year book of Nepal, 2007. Central Bureau of Statistics (CBS), Kathmandu

CBS (2011) Poverty in Nepal 2010/11. Central Bureau of Statistics (CBS), Kathmandu

Dhakal S (2001) Climate change initiatives and Nepal. Tech J 2001(1):24–31

DoF (2007) Current data on pro-poor leasehold forestry programme. Hamro Kalpabriksha, 18(197):23–24

EIA (2009) Nepal energy profile. Retrieved 1 December 2010, from US Energy Information Administration (EIA): http://tonto.eia.doe.gov/country/country_time_series.cfm?fips=NP

FAO (2000) Group promotors effectively raise awareness of poor leasehold forestry farmers in Nepal. Poverty Alleviation Initiat 10(3):13–15

FAO (2006) Global forest resources assessment 2005: progress towards sustainable forest management. Food and Agriculture Organization of the United Nations, Rome

Gautam D, Gautam K, Poudel D (2007a) Climate change impacts and adaptation strategies by poor and excluded communities in Western Nepal: a comparative study of Banganga river basin; Arghakhanchi and Kapilvastu, Nepal. ActionAid Nepal, Kathmandu

Gautam D, Jnavaly S, Sharma A, Amatya A (2007b) Climate change adaptation on livelihood of women farmers: case study of Banke and Bardiya districts of Nepal. ActionAid Nepal, Kathmandu

Gum W, Singh PM, Emmett B (2009) Even the Himalayas have stopped smiling: climate change, poverty and adaptation in Nepal. Oxfam International, Lalitpur

ICIMOD & UNEP (2001) Inventory of glaciers, glacier lakes and glacial lake outburst floods, monitoring and early warning systems in the Hindu Kush Himalayan region. International Centre for Integrated Mountain Development (ICIMOD) & United Nations Environment Programme (UNEP), Kathmandu

ICIMOD (2011) ICIMOD pilots first Forest Carbon Trust Fund in Nepal—Helps communities benefit from forest conservation and sustainable use. Retrieved 16 June 2011, from ICIMOD: www.icimod.org/?q=3743

IPCC (2001) Climate change 2001: the scientific basis. Cambridge University Press, Cambridge

IPCC (2007a) Climate change 2007: synthesis report. Working group contribution to the Fourth assessment report of Intergovernmental Panel on Climate Change. Intergovernmental Panel on Climate Change (IPCC), Geneva

IPCC (2007b) Climate change 2007: the physical science basis. Contribution of Working Group I to the Fourth Assessment Report of the Intergovernmental Panel on Climate Change. Cambridge University Press, Cambridge

IPCC (2007c) Climate change 2007: mitigation of climate change. Contribution of Working Group III to the Fourth Assessment Report of the Intergovernmental Panel on Climate Change. Cambridge University Press, Cambridge
IPCC (2007d) Climate change 2007: impacts, adaptation, and vulnerability. Contribution of Working Group II to the Fourth Assessment Report of the Intergovernmental Panel on Climate Change. Cambridge University Press, Cambridge
Joshi NP (2011) Poverty and climate change in Nepal: poverty analysis in Far-Western Rural Hills. Unpublished PhD dissertation. Hiroshima University, Hiroshima
Joshi NP, Maharjan KL, Piya L (2010) Poverty and food insecurity in Nepal: a review. J Int Dev Coop 16(2):1–19
Joshi NP, Maharjan KL, Piya L (2011a) Effect of climate variables on yield of major food crops in Nepal: a time-series analysis. J Contemp India Stud Space Soc 1:19–26
Joshi NP, Maharjan KL, Piya L (2011b) Effect of climate variables on yield of Rice in Nepal: a panel data analysis with regional disaggregation. In: 7th Asian Society of Agricultural Economists International Conference. Asian Society of Agricultural Economist, Hanoi
Karky BS, Banskota K (2009) Reducing emissions from Nepal's community managed forests: discussion for COP 14 in Poznan. J For Livelihood 8(1):43–47
Koch-Mathian S (2010) Case study 4: Biogas Support Programme (BSP)—Nepal. Sustainable Architecture and Energy Projects Scaling Up (SAESUp), Lyon
Laudari R (2008) Biogas support program of Nepal: current situation and experiences/lesson learnt. Regional forum on bio-energy sector development: challenges, opportunities and the way forward. United Nations Economics and Social Commission for Asian and the Pacific & Asian and Pacific Center for Agricultural Engineering and Machinery, Bangkok. Retrieved 9 February 2010, from http://www.unapcaem.org/Activities%20Files/A0801/0207.pdf
Malla G (2003) Impact of climate change on water and soil health. Agric Environ 63–71
Malla G (2008) Climate change and its impact on Nepalese agriculture. J Agric Environ 9:62–71
Maplecroft (2010) Big economies of the future—Bangladesh, India, Philippines, Vietnam and Pakistan—most at risk from climate change. Retrieved 9 March 2011, from Maplecroft: risk, responsibillity and reputation: http://maplecroft.com/about/news/ccvi.html
MoAC (2005) Statistical information on Nepalese agriculture: time-series information. Agri-Business Promotion and Statistics Division, Ministry of Agriculture and Cooperation (MoAC), Kathmandu
MoAC (2006) Statistical information on Nepalese agriculture. Agri-Business Promotion and Statistics Division, Ministry of Agriculture and Cooperatives (MoAC), Kathmandu
MoE (2010) Developing Nepal's strategic program for climate resilience: prioritisation planning process. Ministry of Environment (MoE), Kathmandu
MoE (2011) Climate change policy, 2011. Ministry of Environment (MoE), Government of Nepal, Kathmandu
MoPE & UNEP (2004) Initial national communication to conference of the parties of the United Nations Framework Convention on Climate Change. Ministry of Population and Environment (MoPE) & United Nations Environment Programme UNEP, Kathmandu
NPC (2003) The tenth plan (Poverty Reduction Strategy Paper) 2002–2007. National Planning Commission (NPC), Kathmandu
NPC (2007) Three years interim plan 2007/08–2009/10. National Planning Commission (NPC), Kathmandu
Olivier JG, Peters JA (2010) No growth in the total global CO_2 emissions in 2009. Netherlands Environmental Assessment Agency (PBL), Bilthoven
Piya L, Maharjan KL, Joshi NP (2011) Forest and food security of indigenous people: a cse of Chepangs in Nepal. J Int Dev Coop 17(1):113–135
Pokhrel AP (2007) Impact of climate change on water resources in Nepal. In: National workshop on climate change and human health: potential impact, vulnerability and adaptation in Nepal. Nepal Health Research Council, Kathmandu
Rana B, Shrestha AB, Reynolds JM, Aryal R, Pokhrel AP, Budhathoki KP (2000) Hazard assessment of the Tsho Rolpa glacial lake and ongoing remediation measures. J Nepal Geol Soc 22:563–570

Regmi B, Adhikari A (2007) Human development report 2007: climate change and human development—risk and vulnerability in warming world; Country case study-Nepal. United Nations Development Programme, New York

Shrestha AB, Wake CP, Dibb JE, Mayewski PA (2000) Precipitation fluctuations in the Nepal Himalaya and its vicinity and relationship with some large scale climatological parameters. Int J Climatol 20:317–327

UNDP (2007) Millennium Development Goals indicators. Retrieved 9 March 2010, from United Nations Development Programme: http://mdgs.un.org/unsd/mdg/Data.aspx

UNEP (n.d.). Company profile: love green Nepal. Retrieved 11 June 2011, from CDM Bazaar: www.cdmbazaar.net/repo/sellers/seller-827688724

Uphoff N (2007) The System of Rice Intensification (SRI) as a system of agricultural innovation. In: Farmer first revisited: Farmer participatory research and development twenty years on. Institute of Development Studies, Sussex

Upreti R (2008) System of Rice Intensification (SRI) experience of Nepal. In: Third national symposium on SRI, SRI India, Tamilnadu

Vidal J (2006) Nepal's farmers on the frontline of global climate change: Himalayan communities face catastrophic floods as weather pattern alter. Retrieved from The Guardian: http://www.guardian.co.uk/environment/2006/dec/02/christmasappeal2006.frontpagenews

WB (2007a) Climate variability, climate change, and adaptation: an emerging framework for engagement in Nepal. Identifying priorities and opportunities of action on climate change adaptation in Nepal. Ministry of Environment, Sceince and Technology, World Bank, Department of Hydrology and Meteorology, and Winrock International, Kathmandu

WB (2007b) Nepal: micro-hydro promotion by Alternative Energy Pormotion Centre (AEPC). Retrieved 2 June 2011, from World Bank Carbon Finance: http://wbcarbonfinance.org/Router.cfm?Page=CDCF&FID=9709&ItemID=9709&ft=Project&ProjID=33540

WB (2008) Towards a strategic framework on climate change and development for the World Bank group; Concept and issue paper (Consultation draft). The World Bank (WB), Washington, DC

WFP (2010) Nepal food security bulletin (Issue 26, October–December, 2009). World Food Programme (WFP), Kathmandu

WHO (n.d.) Malaria endemic districts of Nepal. Retrieved 1 June 2010, from World Health Organization (WHO): http://www.searo.who.int/LinkFiles/Malaria_in_the_SEAR_NEp_ende_dist.pdf

Chapter 9
Effect of Climate Variables on Yield of Major Food-Crops in Nepal: A Time-Series Analysis

Abstract Therefore, this chapter assesses the effect of observed climate variables on yield of major food-crops in Nepal, namely rice, wheat, maize, millet, barley and potato based on a regression model for historical (1978–2008) climatic data and yield data for the food-crops. The yield growth rate of all the food-crops is positive. However, the growth rate for all crops, except potato and wheat, is below the population growth rate during the period. Climate variables like temperature and precipitation are the important determinants of crop yields. Trend of precipitation is neither increasing nor decreasing significantly during this period. However, the temperature is increasing by 0.7 °C during the period. Climate variables show some influences on the yield of these major food-crops in Nepal. Increase in the summer rain and maximum temperature has contributed positively to rice yield. Also, increase in the summer rain and minimum temperature has a positive impact on potato yield. However, increase in the summer rain and maximum temperature adversely affected the yield of maize and millet. Increase in wheat and barley yield is contributed by the current trend of winter rain and temperature. Consideration of spatial variation in similar type of study in Nepal that will be helpful in identifying the region more vulnerable to climate change in terms of crop yield is highly recommended.

Keywords Maize • Precipitation • Rice • Temperature • Wheat

This chapter draws from PhD dissertation of Niraj Prakash JOSHI (2011). Findings discussed in this chapter was presented in the 60th Annual Meeting of the Association of Regional Agriculture and Forestry Economics (ARFE), October 22nd to 24th 2010, Kyoto University, Kyoto, Japan, and published in Journal of Contemporary India Studies: Space and Society, Volume 1, 19–26.

K.L. Maharjan and N.P. Joshi, *Climate Change, Agriculture and Rural Livelihoods in Developing Countries*, Advances in Asian Human-Environmental Research,
DOI 10.1007/978-4-431-54343-5_9,

9.1 Introduction

Nepal is a developing country with a majority (63.7 %) of its population living under the poverty (Alkire and Santos 2010; Joshi et al. 2010). Huge proportions of its population derive their livelihoods from farm and forest, which are highly dependent on natural phenomena. Besides, the disaster prone nature of the country due to its rugged terrain, steep topography, and fragile geological conditions places Nepal among the countries having a high degree of vulnerability to climate change. Such high degrees of vulnerability pose threats to water resources, agriculture, forestry and biodiversity, and human health (Maharjan et al. 2009). Agriculture has been a major concern in the discussions on climate change as food production is essential for sustaining and enhancing human welfare (McCarl et al. 2001; Schmidhuber and Tubiello 2007).

Climate is a primary determinant of agricultural productivity especially in the case of developing countries like Nepal where agriculture is basically dependent on natural circumstances against the controlled environmental condition in developed countries. Therefore, climate change would influence crop yield, thereby crop production to a greater extent in developing countries. Plausible scenarios in climate change i.e., increasing temperature, changes in precipitation, climate extremes like drought, flood and landslides, and higher CO_2 concentrations will directly affect crop yields. In general, the temperature increase will reduce yields and quality of food-crops thereby exacerbating vulnerability in food supply. Similarly, changes in precipitation patterns i.e., intensive rain concentrated in a particular month has a devastating effect on crop production (Abrol and Ingram 1996; Adams et al. 1998; McCarl et al. 2001).

Despite such a high degree of vulnerability to climate change for agriculture vis-à-vis welfare in developing countries, there are limited researches conducted in case of developing countries (You et al. 2005; Mendelsohn 2009; Boubacar 2010; Holst et al. 2010) and very few in the case of Nepal (Malla 2008) The vast majority of such researches are done in developed countries (Stooksbury and Michaels 1994; Lobell and Asner 2003; Chi-Chung et al. 2004; Carew et al. 2009). There are very limited literatures in the case of developing countries, which are going to be adversely affected by predicted climate change (Stooksbury and Michaels 1994; Lobell and Asner 2003; Chi-Chung et al. 2004; IPCC 2007). Therefore, this study analyzes the effect of climate variables on yield of major food-crops in Nepal based on the historical data. An understanding of the national impacts of recent climate trends on major food-crops would help to anticipate the impacts of future climate changes on food self-sufficiency of the country.

9.2 Methodology

Temperature, precipitation, and solar radiation are the three most widely used climate variables to assess climate change and its impact. However, solar radiation has a close positive correlation with maximum temperature. In general, higher solar radiation leads to a higher maximum temperature and lower solar radiation leads to a lower minimum temperature because of radiative cooling (Peng et al. 2004). This shows the direct correlation between temperature and solar radiation. Therefore, to overcome the possible correlation among the independent variables, this study considers only temperature and precipitation. Rainfall is the most important form of precipitation in terms of meeting water requirement of agricultural crops. Daily mean air temperature is the widely used temperature variable to assess the effects of global warming on grain yield. The use of mean air temperature assumes no difference in the influence of day versus night temperature. However, the inclusion of minimum and maximum temperature in the assessment will capture differential effects of day and night temperature (Peng et al. 2004) as well as climate extremities to some extent.

Simulation models and regression models are widely used to estimate the effects of environmental changes on crop productivity levels. Most studies on the possible impact of climate change on crop yields used mainly indirect crop simulation models that make use of crop biophysical simulation. There are relatively limited studies based on regression models (Peng et al. 2004; You et al. 2005; Isik and Devadoss 2006; Mendelsohn 2009; Boubacar 2010). Crop simulation type of study will help to understand the physiological effects of high temperature on crop yield but not the effects of a small increase in temperature associated with global warming (Schlenker and Roberts 2008). In addition, though it is unequivocal that global warming is inevitable in the coming century, even if emissions of greenhouse gases are stabilized at the current level, there exists debate and uncertainty on the extent of warming as well as other related changes (IPCC 2007; Rosegrant et al. 2008). Thus, predictions of the yield changes in response to changes in climate variables, from regression models based on historical climatic data and yield data for specific crops are relatively accurate (Mendelsohn et al. 1994; Lobell and Asner 2003; Lobell et al. 2005; Lobell and Field 2007; Boubacar 2010). This can be done through application of production function as follows (Nicholls 1997; Lobell and Field 2007);

$$\Delta Yield = m + r_y \Delta Climate + \varepsilon$$

Here,

ΔYield is the observed trend in yield, m is the average yield change due to management and other non-climatic factors (e.g. increased CO_2), ΔClimate is the observed trend in temperature and rainfall, r_y is the yield response to this trend, and ε is the residual error.

Detrending of the yield and climate variables and using the residuals to calculate quantitative relationships between variation in climate and yield can remove non-climatic influences such as adoption of new cultivars and changes in crop management practices (Nicholls 1997; Lobell and Field 2007). Detrending can be done by using the first-difference time-series for yield and climate variables i.e., the difference in values from 1 year to the next.

Paddy, maize, millet, wheat, barley, and potato are the major food-crops of Nepal as these crops are used to meet the basic food requirement of its population (Subedi 2003). Paddy, maize, potato and millet are the main food-crops cultivated during the summer season (from May to August), whereas wheat and barley are the main

Jan	Feb	Mar	April	May	June	July	Aug	Sept	Oct	Nov	Dec	Season
Mountain (Rainfed)												
		Mai-P	Mai-P				Mai-H	Mai-H	Mai-H			Summer
			Mil-P	Mil-P					Mil-H	Mil-H		Summer
	Pot-P	Pot-P				Pot-H	Pot-H	Pot-H	Pot-H			Summer
				Whe-H	Whe-H					Whe-P	Whe-P	Winter
			Bar-H	Bar-H						Bar-P	Bar-P	Winter
Hills (Partial irrigation/Rainfed)												
				Pad-TP	Pad-TP			Pad-H	Pad-H			Summer
		Mai-P	Mai-P				Mai-H	Mai-H				Summer
					Mil-P	Mil-P			Mil-H	Mil-H		Summer
		Pot-P	Pot-P				Pot-H	Pot-H				Summer
		Whe-H	Whe-H	Whe-H					Whe-P	Whe-P	Whe-P	Winter
		Bar-H	Bar-H						Bar-P	Bar-P	Bar-P	Winter
Hill (Irrigated)												
		Pad-TP	Pad-TP			Pad-H	Pad-H					Spring
	Mai-P	Mai-P			Mai-H	Mai-H						Spring
		Whe-H	Whe-H	Whe-H					Whe-P	Whe-P	Whe-P	Winter
Tarai (Rainfed)												
					Pad-TP	Pad-Tp		Pad-H	Pad-H	Pad-H		Summer
			Mai-P	Mai-P			Mai-H	Mai-H				Summer
		Whe-H	Whe-H						Whe-P	Whe-P		Winter
Tarai (Irrigated)												
						Pad-Tp	Pad-Tp			Pad-H	Pad-H	Late-summer
	Mai-P	Mai-P			Mai-H	Mai-H						Spring
		Pad-Tp	Pad-Tp			Pad-H	Pad-H	Pad-H				Spring
	Mai-H	Mai-H							Mai-P	Mai-P		Winter

Mai-Maize, Mil-Millet, Whe-Wheat, Bar-Barley, Pad-Paddy, P-Plantation, Tp-Transplantation, H-Harvesting

Plantation/transplantation Crop growing phase Harvesting

Fig. 9.1 Crop calendars of major food-crops in Nepal (FAO and WFP 2007)

winter crops cultivated from November to February. Figure 9.1 shows the crop calendar of these major food-crops in Nepal.

Due to the consistency in the availability of climate data from the maximum number of stations existing in the country, the period from 1978 to 2008 is taken into consideration. A period of more than 30 years is qualified for study of the impact of climate variables on the yield of the food crops as response to climate change (IPCC 2007). Average national yields of the food-crops for 1978–2008 were compiled from different publications of the Ministry of Agricultural and Cooperatives. Similarly, the crude data of climate variables, i.e. temperature and rainfall were obtained from Department of Hydrology and Meteorology, Nepal, on a monthly basis from 1977 to 2008. Rainfall data from 235 weather stations distributed along the elevation from 72 masl to 3,803 masl, and temperature data from 45 stations distributed along the elevation from 72 to 2,680 masl were compiled for the purpose of this study. Rather than using annual averages for each climatic variable, we defined an effective growing season for each crop based on the contiguous months within the growing season for major ecological regions.

9.3 Results and Discussion

9.3.1 Trend of Food-Crops' Yield

The yield trend of the food-crops based on the regression coefficient against time shows that time has significant (P-value < 0.00) effect on yield of all the food-crops.

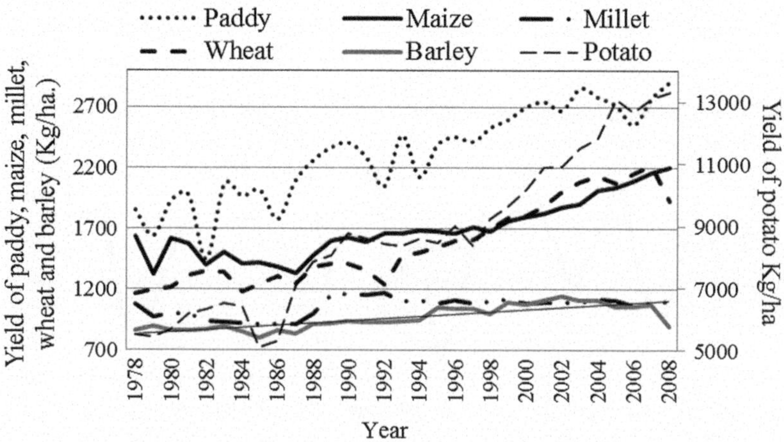

Fig. 9.2 Yield trends of major food-crops (MoAC 2009; MoA 1990)

However, the trend of yields for the six major food-crops shows very different patterns (Fig. 9.1). Potatoes have the highest regression coefficient against time variable. Yield of potato is growing by 0.26 ton/ha every year. Thus, yield of potato has increased from 5.5 ton/ha in 1978 to 13.3 ton/ha in 2008 contributing the yield growth rate of 3.32 %. Except for the year 1985, during which the yield of potato declined sharply, potato yield has been continuously increasing. There is no relation with climate variables for such sharp decline in yield. Wheat also shows better performance in terms of yield growth. With the regression coefficient of 0.035 against the time variable, the yield growth rate of wheat is 2.32 %. Yield growth rate of only these two crops is higher compared to population growth rate (2.3 %) of the country.

Yields of paddy and maize are also growing but the growth rate is well below the population growth rate. They are growing at the rate of 1.7 % and 1.49 %, respectively. A sharp decline in the yield of paddy and maize in 1982 can be linked to sharp decline in summer rain in the same year. Yield decline in paddy and maize is directly associated with summer rain (Figs. 9.2 and 9.3). Yield growth of barley and millet, which are also a minor food-crops are relatively stagnant, growing at the rate of below 1 %.

9.3.2 *Trend of Climate Variables*

Trend of climate variables are analyzed on a seasonal basis to coincide the growing seasons of the crops considered for the study. Average of temperature for the effective growing season based on the data obtained from all meteorological stations is taken into consideration, whereas in case of rainfall, average monthly rainfall of the months added to get total rainfall for the season is considered. Accordingly, trends of the minimum temperature, maximum temperature, and rainfall for summer and

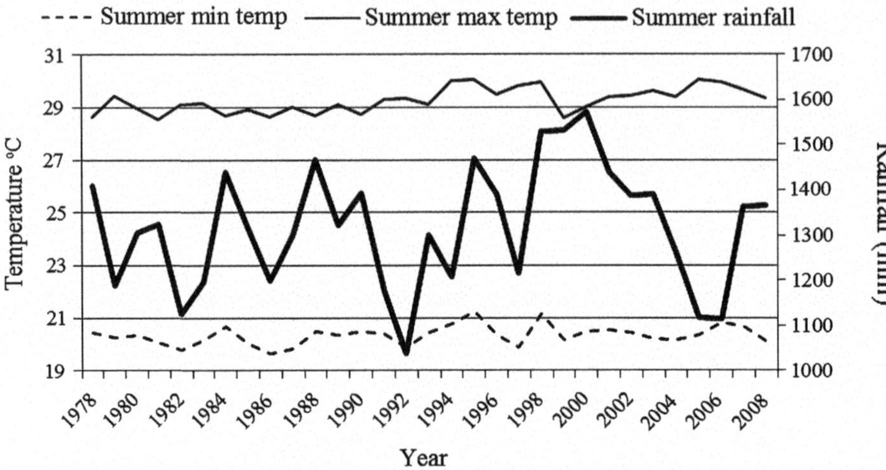

Fig. 9.3 Trend of total summer rainfall, and average summer minimum and maximum temperature (raw data from Department of Hydrology and Meteorology (DHM), Nepal); *Note*: Summer season includes the months of May, June, July and August

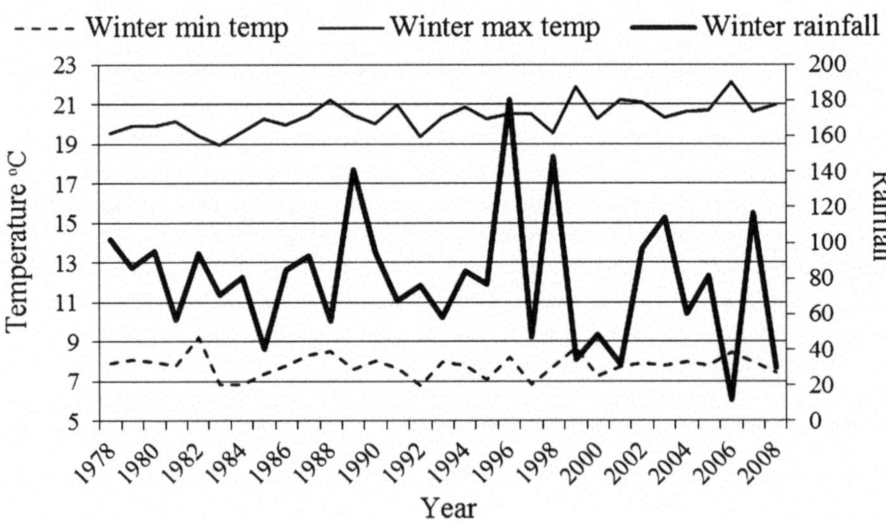

Fig. 9.4 Trend of total winter rainfall, and average winter minimum and maximum temperature (raw data from DHM, Nepal). *Note*: Winter season includes the months of November and December of the preceding year and January, and February of succeeding year

winter are presented in Figs. 9.3 and 9.4, respectively. Here, only a maximum temperature in winter and summer season shows significant (P-value < 0.00) increase over time, whereas the minimum temperatures and rainfall for both seasons show insignificant associations with a time variable.

Table 9.1 Relationship between yield of summer food-crops and summer climate variables

	Paddy		Maize		Millet		Potato	
Variable	Coefficient	*P*-value	Coefficient	*P*-value	Coefficient	*P*-value	Coefficient	*P*-value
Sumrain	0.012 (0.004)	0.01***	−0.002 (0.002)	0.43	−0.001 (0.001)	0.56	0.003 (0.013)	0.79
Summintemp	−0.15 (0.12)	0.25	0.07 (0.06)	0.32	0.02 (0.03)	0.55	0.21 (0.38)	0.58
Summaxtemp	0.06 (0.11)	0.61	−0.13 (0.06)	0.04**	−0.04 (0.03)	0.19	−0.02 (0.35)	0.95
R^2	0.40		0.18		0.08		0.07	

Note:
**Significant at the 0.05 level
***Significant at the 0.01 level
Sumrain summer rainfall, *Summintemp* summer minimum temperature, *Summaxtemp* summer maximum temperature, figures in parentheses indicates standard error

Rainfall fluctuates over the years with less degree of predictability. However, it is in increasing trend for the summer season, but in decreasing trend for winter. The coefficients suggest that summer rainfall is increasing by 2.2 mm every year, whereas winter rainfall is decreasing by 0.63 mm every year. Rainfall in Nepal is concentrated in summer. Around 75 % of rainfall occurs during this season. The positive coefficient for summer rainfall and negative coefficient for winter rainfall indicates that rain in the summer is becoming more intense, which could hamper yield of summer food-crops due to water borne disaster like flood and landslides. However, still the relationship between rainfall and yield show positive correlation, i.e. yield will grow with increased rainfall and shrink with decreased rainfall.

Coefficients of temperature for both seasons are positive except for winter minimum temperature. The winter maximum temperature is increasing at a higher rate compared to summer maximum temperature. Summer and winter maximum temperature is increasing at the rate of 0.03 °C and 0.05 °C each year between 1978 and 2008, respectively. Summer minimum temperature is also increasing every year by 0.01 °C. However, the winter minimum temperature is decreasing each year but at a very low rate 0.001 °C every year. Increase in temperature up to 2 °C will increase the food-crops yields in Nepal (Malla 2008). Therefore, the increase in temperature during the period from 1978 to 2008 i.e. below 2 °C would be favorable for growth in yield of food-crops. However, a decline in minimum winter temperatures could hamper the yield of winter crops as frost frequency caused by the decline in minimum winter temperatures influence wheat yield adversely (Nicholls 1997).

9.3.3 *Climate Yield Relationships*

Multivariate regression analysis of the first difference in yield of the crops considered for this study is presented separately for both summer crops and winter crops in Tables 9.1 and 9.2, respectively. The results suggest that the model is able to describe a variation in food-crops yield ranging from 40 % in the case of paddy to only 2 % in the case of barley. Though, the regression results show very few significant relationships between yield and climate variables, such coefficient can be used to assess real effect of climate variables in change of yield of food-crops considered

Table 9.2 Relationship between yield of winter food-crops and winter climate variables

	Wheat		Barley	
Variable	Coefficient	*P*-value	Coefficient	*P*-value
Winrain	0.003 (0.004)	0.52	0.001 (0.002)	0.69
Winmintemp	0.04 (0.03)	0.15	−0.002 (0.02)	0.87
Winmaxtemp	0.008 (0.03)	0.78	0.01 (0.02)	0.54
R^2	0.17		0.02	

Note: *Winrain* winter rainfall, *Winmintemp* winter minimum temperature, *Winmaxtemp* winter maximum temperature

Table 9.3 Change in yield of food crops due to current climate trends in Nepal

Crops	β_1	ΔR	β_2	ΔT_{min}	β_3	ΔT_{max}	ΔYield (kg/ha)
Paddy	0.012	−4.31	−0.15	−0.34	0.06	0.70	40.82
Maize	−0.002	−4.31	0.07	−0.34	−0.13	0.70	−106.04
Millet	−0.001	−4.31	0.02	−0.34	−0.04	0.70	−30.45
Potato	0.003	−4.31	0.21	−0.34	−0.02	0.70	−97.67
Wheat	0.003	14.25	0.04	−0.48	0.008	1.45	35.09
Barley	0.001	14.25	0.002	−0.48	0.01	1.45	27.80

Note: β_1 coefficient of rainfall, ΔR change in rainfall, β_2 coefficient of minimum temperature, ΔT_{min} change in minimum temperature, β_3 coefficient of maximum temperature, ΔT_{max} change in maximum temperature, ΔYield change in yield [$=(\boldsymbol{\beta_1 * \Delta R}) + (\boldsymbol{\beta_2 * \Delta T min}) + (\boldsymbol{\beta_3 + \Delta T max})$]

for this study (Nicholls 1997). In addition, a sign of coefficients gives the direction of the yield movements against changes in climate variables. Climate variables show significant relations with paddy and maize only. The coefficient indicates that paddy yield increases significantly with the increase in summer rainfall. Maize yield shows a negative relation with summer maximum temperatures, i.e., if the summer maximum temperature increases the yield of maize will decline sharply.

9.3.4 *Change in Yield Due to Climate Trend*

Change in food-crops' yield due to climate variables is calculated using coefficient of the climate variables for the respective crops and observed changes in the climate variables during the study period i.e., $\Delta Y_i = (\beta_{1i} * \Delta R) + (\beta_{2i} * \Delta T_{min}) + (\beta_{3i} + \Delta T_{max})$. Here, ΔY_i is observed change in yield of ith crop due to climate variable, and β_{1i}, β_{2i}, and β_{3i}, are coefficient of rainfall, maximum summer temperature, and minimum summer temperature, respectively for i[th] crop. Similarly, ΔR, ΔT_{min}, and ΔT_{max} are observed changes in rainfall, summer minimum temperature, and summer maximum temperature, respectively during the study period.

The current trend in climate variables has contributed positively to yield of both winter crops namely; wheat and barley. In the case of wheat, there is a 814 kg increase of yield during the study period, out of which 35.1 kg is contributed by the current climate trend (Table 9.3). Here, decreasing winter rain and winter minimum temperature offset the positive effect of increased winter maximum temperature.

For barley, the current climate trend contributed around 50 % of the yield increase. Such increase can be attributed to increased winter maximum temperature and decreased winter minimum temperature. In the case of summer crops, only paddy is favored by the current climate trend. It has contributed 41 kg increase in yield in case of paddy. An increase in the summer rain and summer maximum temperatures has contributed highly in such increase. Other crops, especially maize, are adversely affected by the current climate trend in Nepal. The adverse impact of increased summer maximum temperature and summer rain are the main factors, which caused suppression of yield by 106 kg/ha, and 30 kg/ha for maize and millet, respectively. In the case of potatoes, it is the adverse impact caused by the increase in summer maximum temperature that offsets the positive impact of increased summer rain and summer minimum temperature. Here, the current climate trends suppress the yield of potatoes by 98 kg/ha.

9.4 Conclusion

This paper analyzed the impact of current climate trends on yields of six main food-crops in Nepal. These food-crops are divided into two groups based on their growing season, namely; summer and winter season crops. The impact is assessed for each crop based on the growing season of respective crop. Yield of potatoes, wheat, paddy, and maize is in a growing trend, but fluctuates over the years, whereas the yield of millet and barley, two minor cereal crops, is growing very steadily. In summer, each of the climate variables is in increasing trend, whereas in winter, rainfall and minimum temperature is decreasing. In summer, increase in rain and maximum temperature has contributed positively to yield growth of paddy. Similarly, increase in wheat and barley yield is contributed by current climate trends. However, increased summer rain and maximum temperature suppressed the yield growth of maize and millet, whereas the negative impact of increase in the summer maximum temperatures outweighed the positive impacts of increased summer rain and summer minimum temperatures in the case of potatoes.

This study, thus, concludes that food-crops grown in summer are adversely affected by the current trend of climate. Except for paddy, which has high water demand and thrives on water logging condition, other summer crops are adversely affected by an increase in rainfall and maximum temperature. On the other hand, though rainfall is at a declining trend in winter, increase in temperature has positively contributed to the yield growth of both winter crops. With this, we can recommend that any program dealing with minimizing the adverse impact of climate change on food-crops production should first consider the crops like maize and potato, which are being affected at higher degree compared to other food-crops. Moreover, these two crops are an important staple food in case of Nepal especially in Mountain and Hills that are also exposed to a higher degree of vulnerability to climate change. The main shortcoming of this study is treating the whole country as one basket despite the huge diversity existing within. Therefore, it is highly recommended to conduct similar studies considering the variation caused by ecological and administrative division of the country.

References

Abrol YP, Ingram KT (1996) Effects of higher day and night temperatures on growth and yields of some crop plants. In: Bazzaz F, Sombroek W (eds) Global climate change and agricultural production: direct and indirect effects of changing hydrological, pedological and plant physiological processes. Wiley, West Sussex, pp 123–140

Adams RM, Hurd BH, Lenhart S, Leary N (1998) Effects of global climate change on agriculture: an interpretative review. Clim Res 11:19–30

Alkire S, Santos ME (2010) Nepal: country briefing. Acesso em 8 de August de 2010, disponível em Oxford Poverty and Human Development Initiative (OPHI) Multidimensional Poverty Index Country Briefing Series: http://www.ophi.org.uk/wp-content/uploads/Nepal.pdf

Boubacar I (2010) The effects of drought on crop yields and yield variability in Sahel. In: The Southern Agricultural Economics Association annual meeting. The Southern Agricultural Economics Association, Orlando

Carew R, Smith EG, Grant C (2009) Factors influencing wheat yield and variability: evidence from Manitoba, Canada. J Agric Appl Econ 41(3):625–639

Chi-Chung C, McCarl BA, Schimmelpfennig DE (2004) Yield variability as influenced by climate: a statistical investigation. Clim Change 66:239–261

FAO & WFP (2007) FAO/WFP food security assessment to Nepal (special report). Food and Agriculture Organization of the United Nations (FAO) and World Food Programme (WFP), Rome

Holst R, Yu X, Grun C (2010) Climate change, risk and grain production in China. In: 2010 AAEA, CAES, WAEA joint annual meeting. Agriculture and Applied Economics Association, Colorado, p 29

IPCC (2007) Climate change 2007: synthesis report. Working group contribution to the Fourth assessment report of Intergovernmental Panel on Climate Change. Intergovernmental Panel on Climate Change (IPCC), Geneva

Isik M, Devadoss S (2006) An analysis of the impact of climate change on crop yields and yield variability. Appl Econ 38:835–844

Joshi NP, Maharjan KL, Piya L (2010) Poverty and food insecurity in Nepal: a review. J Int Dev Coop 16(2):1–19

Lobell DB, Asner GP (2003) Climate and management contributions to recent trends in U.S. agricultural yields. Science 299(5609):1032

Lobell DB, Field CB (2007) Global scale climate-crop yield relationships and the impacts of recent warming. Environ Res Lett 2(1):1–7

Lobell DB, Ortiz-Monasterio JI, Asner GP, Matson PA, Naylor RL, Falcon WP (2005) Analysis of wheat yield and climatic trends in Mexico. Field Crop Res 94:250–256

Maharjan KL, Joshi NP, Piya L (2009) Sources of climate change, its impact, and mitigation issues in Nepal. In: International conference on climate change, livelihoods, and food security. Institute of Development Studies and Association of Asian Scholars, Jaipur

Malla G (2008) Climate change and its impact on Nepalese agriculture. J Agric Environ 9:62–71

McCarl BA, Adams RM, Hurd BH (2001) Global climate change and its impact on agriculture. Texas A&M University, Texas

Mendelsohn R (2009) The impact of climate change on agriculture in developing countries. J Nat Resour Policy Res 1(1):5–19

Mendelsohn R, Nordhaus WD, Shaw D (1994) The impact of global warming on agriculture: a Ricardian analysis. Am Econ Rev 84(4):753–771

MoA (1990) Agricultural statistics of Nepal: 1990. Agricultural Statistics Division, Ministry of Agriculture (MoA), Lalitpur

MoAC (2009) Statistical information on Nepalese agriculture 2008/2009 (2065/066). Agri-Business Promotion and Statistics Division, Ministry of Agriculture and Cooperatives (MoAC), Kathmandu

Nicholls N (1997) Increased Australian wheat yield due to recent climate trends. Nature 387: 484–485
Peng S, Huang J, Sheehy JE, Laza RC, Visperas RM, Zhong X et al (2004) Rice yields decline with higher night temperature from global warming. Proc Natl Acad Sci 101(27):9971–9975
Rosegrant MW, Ewing M, Yohe G, Ian Burton SH, Valmonte-Santos R (2008) Climate change and agriculture: threats and opportunities. German Technical Cooperation (GTZ), Eschborn
Schlenker W, Roberts MJ (2008) Estimating the impact of climate change on crop yields: the importance of non-linear temperature effects. NBER Working Paper 13799. National Bureau of Economic Research, Massachusetts
Schmidhuber J, Tubiello FN (2007) Global food security under climate change. Proc Natl Acad Sci 104(50):19703–19708
Stooksbury DE, Michaels PJ (1994) Climate change and large-area corn yield in the Southeastern United States. Agron J 86(3):564–569
Subedi BP (2003) Population and environment: a situation analysis of population, cultivated land and basic crop production in Nepal in 2001. In: Population monograph of Nepal, vol. II. Central Bureau of Statistics, Kathmandu, pp 1–35
You L, Rosegrant MW, Fang C, Wood S (2005) Impact of global warming on Chinese wheat productivity. International Food Policy Research Institute (IFPRI), Washington, DC

Chapter 10
Community Perceptions of Climate Change and Its Impacts

Abstract This chapter analyzes the community perception of climate change from international to national perspectives. In the later part it assesses the community perception of climate change, triangulation of such perceptions with meteorological data and impacts of such perceptions on the livelihoods of rural people. The analysis is based on the preliminary data generated through household surveys particularly focused on highly marginalized indigenous nationalities of Nepal residing in the remote Mid-hills of Nepal.

Keywords Chepang • Climate change impacts • Community perception • Nepal mid-hills • Public awareness

10.1 Introduction

Community perceptions, views, and opinions regarding climate change matters both in designing mitigation policies as well as formulating adaptation strategies. According to Leiserowitz (2007), public opinion is important because it forms the background within which policies are formulated, and policies are supported or opposed by the public based on how they perceive the associated risks of climate change. If the public perception of risks differs from the view of policy makers, policy implementation will be misunderstood, neglected, or even opposed (Lorenzoni and Pidgeon 2006). On the mitigation side, measures like limiting the use of fossil fuels and promoting renewable energies might require changes in the energy tax policies, which cannot be implemented unless the act is accepted as necessary and supported by the general public. Thus, public views and opinions must be assessed

---

This chapter draws partially from PhD research of Luni PIYA. Part of findings discussed in this chapter has been submitted for publication in Journal of Contemporary India Studies: Space and Society, Volume 2 (2012): 35–50.

K.L. Maharjan and N.P. Joshi, *Climate Change, Agriculture and Rural Livelihoods in Developing Countries*, Advances in Asian Human-Environmental Research,
DOI 10.1007/978-4-431-54343-5_10, © Springer Japan 2013

and considered by the government before designing and implementing any such mitigation measures. On the other side, the adaptation issue is crucial especially for the vulnerable communities who are more affected by the adverse impacts of climate change. For such communities, how they perceive the ongoing changes determines how they formulate strategies to cope with the changes in the short run and to adapt to the long term changes. In other words, it is necessary to realize that some changes are going on in order to take actions to adjust to those changes (Deressa et al. 2011).

The poorest countries and the poorest communities in any country are the ones who suffer most from the disasters brought about by climate change, because of the lack of means to cope with it. The rural communities all over the world are also the ones to bear the brunt of adverse impacts of climate, mainly because their livelihoods is dominantly dependent upon natural resources based activities, which in turn, are directly impacted by the climate. Monitoring of climate change in the rural areas is often hindered by the lack of data and weather stations. Very often, the experiences and observations of the communities living in these remote rural areas can be the source of information for the scientific communities. It is also important for policy makers and development workers to first assess how such changes in climate patterns are understood by the community. Unless the adaptation options put forth by the organization matches with the perceptions of the community, it cannot be expected that those options will be adopted by the community. It is, thus, important to study the perceptions and views of the general public such that they provide insights in formulating mitigation as well as adaptation strategies.

This chapter discusses the public perceptions and opinions about climate change by reviewing relevant surveys and studies conducted at both national and local levels. The next section of the chapter looks into public opinion about climate change based on the results obtained from the major cross-national polls conducted by some of the research institutions from around the world. The third part deals with the local level community perceptions of climate change, focusing in the rural areas of Africa and Asia. The fourth section presents a case study of community perceptions of a marginalized community in the rural mid-hills of Nepal, followed by the concluding remarks.

10.2 Public Opinions on Climate Change: National and Cross-Country Polls

Public opinion polls on climate change are regularly conducted in developed countries. Some of the institutions conducting such regular polls are the Pew Research Center, Globescan, BBC, and the World Bank. While the Pew and Globescan conduct regular polls in the US, BBC does it for the UK. Meanwhile, these institutes also regularly conduct cross-country surveys that cover US, UK, and most of the EU countries. However, the coverage of Asian and African countries, more specifically developing countries, is much lesser. Most of these polls use the term “global warming” rather than “climate change,” probably because global warming is more common and frequently used among the general public, while climate change is used more among the scientific community. Although the results of these polls might not be directly comparable, owing to the differences in exact phrases used in the questions, the methodologies followed, and different timings and locations of the surveys, these polls do give a broad idea on the opinion and perception of the general public towards "global warming" or "climate change" in the surveyed countries. This section of the chapter makes a review of some representative national and cross-country surveys covering the issues of public awareness about climate change, concern regarding the issue, opinions about the impacts posed by climate change, and attitude about taking the necessary actions, responsibilities and policies to mitigate climate change.

10.2.1 Awareness and Concern About Climate Change

The awareness regarding climate change or global warming is much higher in developed countries compared to the developing countries. As reported by Pew (2006), out of 15 countries surveyed, above 90 % of the respondents in developed countries like the US, the UK, France, Germany, Spain and Japan have heard about the global warming, while the proportion was quite low in developing countries, the lowest being 12 % in Pakistan and 26 % among the Nigerian Muslims. Upon comparing the results from various polls, the level of public awareness of climate change within Asia also shows a similar trend of being proportional to the level of economic development with Japan and South Korea reporting 99 % and 94 % of aware respondents, while the proportion being 57 % in India, and 35 % in Indonesia, and 12 % in Pakistan (Kim 2011). Within the US, a recent poll by Pew (2011) shows that there has been a modest increase in the percentage of people who believe there is solid evidence of global warming over the last 2 years, up from 57 % in 2009 to 68 % in 2011. On the other hand, those in the UK who believe climate change and global warming is happening saw a decrease from 83 % in 2009 to 75 % in 2010 (BBC 2010).

The concern about climate change is found to be increasing among the general public. In a time-series multinational survey conducted by Globescan across 16 countries in 2003 and 2006, it was found that the percentage saying the problem is very serious increased from 49 % to 61 %, with none of the country reporting a decrease. Similar to the level of awareness, the level of concern regarding climate change also shows a similar trend with a higher percentage of aware respondents taking it as a serious problem in developing countries compared to the developed countries. In 2006 a poll of 30 countries was taken, 23 countries had a high response towards global warming being a serious problem, while the remaining seven countries had a lower response. Interestingly, the US (21 %) and China (17 %), the largest emitters, were the ones where the highest percentage of respondents didn't think global warming is a problem (GlobeScan 2006). In a more recent study by the World Bank (2010), it is reported that a vast majority in low-income countries like Bangladesh (85 %), Kenya (75 %), Senegal (72 %), and Vietnam (69 %) thought climate change is a very serious problem. On the other hand, high income countries like US (31 %), Japan (38 %), and France (43 %) had fewer responses saying climate change is a very serious problem, the least being in fast-growing economies of China (28 %) and Russia (30 %). Similarly, in another cross-country poll by Pew in 2006 shows that out of the 15 countries surveyed, the level of worry about global warming was the least in the US (19 %) and China (20 %), while the concern was highest among Japanese (66 %) and Indians (65 %) (Pew 2006). However, the level of concern among the Americans is found to be increasing over the years, with the percentage of respondents taking it as a very serious problem rising to 35 % in 2009 and further to 38 % in 2011 (Pew 2011).

10.2.2 Opinion About the Impacts of Climate Change

In a survey conducted by Globescan in 2001, perception that global warming is a threat to the local community was believed most strongly by the respondents in developing countries like Brazil, Colombia, Kazakhstan, India, Cuba, and Chile, while very few in developed nations like Germany, France, Spain, and US believed so (Leiserowitz 2007). In another study done in the US shows that while 68 % of the respondents were concerned about the impacts on people around the world and on nature, only 13 % of them were concerned about the impacts on themselves, their family and their local community (Leiserowitz 2006). This is an indication that usually residents in developed countries perceive that climate change is not an immediate threat to themselves, and is rather a distant threat. Unless the people begin to experience the impacts themselves, the issues of global warming and climate change will remain a low priority in the developed countries. In 2010, the World Bank asked the respondents in 16 countries whether dealing with climate change should be a priority or not, the majority of those agreeing strongly were from developing countries, the highest being in Vietnam (63 %) followed by Bangladesh (54 %) and Kenya (53 %), while those agreeing strongly were the least in higher-income economies, i.e. the US (14 %), Russia (18 %) and Japan (18 %). In the same survey, the respondents were asked when climate change will affect people substantially, the percentage of respondents saying "now" was highest in Kenya (88 %), Vietnam (86 %), and Mexico (83 %), while those replying "now" were the least in Russia (27 %), the US (34 %) and Egypt (35 %). Interestingly, respondents replying "never" were outstandingly the highest in the US with 14 % followed by Japan with 4 % (WB 2010). This again demonstrates that people in developed countries are less concerned about the negative impacts of climate change. In order to raise concern and priority to climate change among the public, simply providing the correct scientific information is not sufficient; rather Lorenzoni and Pidgeon (2006) opine that situating the climate change "in the locality" is the most important so that the public see the risks posed upon themselves and thus becomes more concerned.

10.2.3 Attitudes Towards Mitigation Action and Related Policies

The World Bank (2010) asked the respondents in 16 countries whether it would be necessary to increase the energy cost so as to encourage energy conservation. The majority of respondents in ten countries said it is necessary, the highest being in Japan (81 %), Kenya (75 %) and Vietnam (70 %). Iran and the US had a divided opinion while the majority disagreed in the countries such as: Russia (81 %),

Mexico (59 %), Brazil (56 %) and France (53 %). Interestingly, the countries which do not support an increase in energy costs are all oil producers except for France. In the same survey, the majority within 14 countries are willing to pay certain amounts of cost, if needed, to support national actions against climate change except for Russia (62 %) and Brazil (59 %), where the majority within said they are not willing to pay any costs to support any national steps against climate change.

In the US, it was found that 90 % thought the country should reduce its emission. However, when it came to taking individual actions, large majorities opposed a gasoline tax (78 %) or a business energy tax (60 %) to reduce greenhouse gas emissions, thereby hinting towards a contradiction in American climate change risk perceptions and policy preferences (Leiserowitz 2006). The similar trend of public opposing policies directly associated with their individual budget is also reported in studies done in the UK in 2002, where 85 % of respondents were willing to change their lifestyles to reduce climate change impacts, and among the available options most (more than 90 %) were ready to follow easy domestic measures like recycling and reusing that didn't cost anything. Whereas, fewer (68 %) favored the option of using cars lesser; furthermore only a few (37 %) were in favor of an increase in the price of petrol. A similar trend is seen among the Asians as well. Kim (2011) prepared a regional summary from the results of various polls done in many countries across the globe. It was found that Asians are least willing to bear the costs of climate change mitigation and also the least supportive of tax incentives for alternative energy development.

These results shed light upon some important behavioral tendencies that need to be addressed upon for mitigating climate change. Although most of the respondents who are aware of global warming and climate change agree that individual contribution is necessary. Although more are willing to do easily affordable domestic activities that would cost nothing like recycling but would not bear the cost or give up luxuries like a private vehicle, paying higher taxes and raising the energy costs. This reveals that there will be an absence of opposition only to those initiatives that are not perceived to have significant impact on individual lifestyle. Furthermore, the majority of people in the developed and highly growing economies disapproves with the necessity to increase energy costs or energy taxes, which suggests a challenge for formulating environmental policies. Responses also show that individuals wish they need not make any sacrifices, thus simply enlisting the measures by which individuals can contribute to mitigate climate change is not enough. Rather there is a need to convince all the community members to do the same thing and make them feel that whole of the society is moving in the same direction. Also, Lorenzoni and Pidgeon (2006) suggest that unless the people should feel that the climate change is impacting not only to distant places but to their locality itself, initiation of behavioral changes is difficult to achieve.

10.3 Perceptions of Climate Change Among Rural Communities

While the national and cross-country polls are important to shed light regarding the general public perceptions of climate change and the support or opposition of the people to certain policies, these surveys have limitations in that they are conducted mainly in the urban areas and are too general to cover the specific aspects at subnational levels and the location specificities of rural communities. Although climate change is a universal phenomenon, its indicators and manifestations are entirely local, so are the adaptation choices, strategies, and practices. There has, thus, been increasing emphasis on the bottom-up approaches that climate change studies should be conducted at the local level where adaptations ultimately take place (Smit and Wandel 2006). There are few studies done among the rural communities in developed countries (Patino and Gauthier 2009). However, many studies have been conducted in the rural localities in Africa (Deressa et al. 2011; Gbetibouo 2009; Maddison 2007) and Asia (Chaudhary and Bawa 2011; Byg and Salick 2009; Vedwan 2006; Dahal 2005; Vedwan and Rhoades 2001). In Asia, all these studies are conducted among the Himalayan communities of India, Nepal, and Tibet, probably because much of the attention in Asia has been received by the melting glaciers in the Himalayas (IPCC 2007). In Nepal, though few other studies on the local perceptions of climate change have been conducted in the hills and low-lying plains (Tiwari et al. 2010; Bhusal 2009), such studies are yet to be circulated widely.

The view of local communities about the ongoing changes in climate, its causes and impacts can be entirely different from what science has explained about climate change. Byg and Salick (2009) report that Tibetans in Yunnan province gave many spiritual reasons like angering of mountain gods as the causes of disruption in the climate patterns. Very often, the understanding of climate change by rural communities is a function of micro-level livelihood practices and is conditioned by the knowledge of crop-climate interaction. For instance, the apple growers in the Northwestern Himalayas of India noticed changes in temperature and rainfall only for the period before apple harvest (Vedwan 2006). Their perceptions of changes in snowfall in the area were very much linked to the various growth stages of apple; like late snowfall was easily noticed by the farmers because the amount of snowfall is very important to determine the fulfillment of chilling requirements to break the winter-dormancy in apples. Similarly, shifts in rainfall hampered the color development of apples and thus was remarkably mentioned by these farmers. However, once the apples are harvested in September, changes in any of the climate variables were rarely reported (Vedwan and Rhoades 2001). It is, thus, very important to first understand how local people understand the climate and how climate interacts with their livelihood activities. Unless adaptation policies and related projects address the local perceptions, it cannot be expected that the community will agree to and adopt the recommended practices. Furthermore, since rural communities are the ones who have closely observed the local climatic patterns, local knowledge can provide important insights into the phenomenon that has not yet been noticed or researched by the scientists. Patino and Gauthier (2009) demonstrate that local perspectives can be combined with scientific climate scenarios to draw policy recommendations from the community through participatory vulnerability mapping.

All the studies cited above show there are some members in every community who do not perceive any changes in climate. Within those who perceive changes, not all of the perceptions match the meteorological records. Attempts have been made to understand the characteristics that differentiate the members who perceive the changes from those who do not perceive any changes across rural communities in Africa (Deressa et al. 2011; Gbetibouo 2009; Maddison 2007). It has been noted that both individualistic and general characteristics affect ability to perceive. While individualistic factors like age and household size do not have much policy implications, others like gender differences, education, and farming experience have important policy relevance. Among other factors, access to information, social networks, infrastructure like distance to market, and engaging in non-farm income sources are found to determine the ability to perceive changes in temperature and rainfall. Furthermore, there are factors specific to agriculture like farm income, farm extension services, nature of farming (subsistence or commercial), soil quality, access to irrigations that affect perception of climate change. Such type of analysis is important as it helps to characterize those members who have the ability to perceive changes in climatic variables versus those who cannot, thereby highlighting the factors that need to be addressed in order to facilitate perceptions and finally adaptations to climate change at the local level. There is a dearth of quantitative studies on the factors determining the community perceptions in Asian context.

Besides the perception of rural communities regarding the trends of climatic variables, studies have also covered the community perceptions about the impacts of such changes. Byg and Salick (2009), in their study conducted in the Tibetan Himalayas report that most of the respondents reported negative impacts of climate change like increasing agricultural pests and diseases, changes in time of planting and harvesting, increased health problems and spoilt food due to higher temperatures; few of the impacts reported are positive like easier to wash and reduced firewood need due to warmer weather. Nominal and ordinal logistic regression of the impacts with age, gender, time spent in the village, and village location showed that the impacts were highly related to the village locations and lower elevation villages showed most of the negative impacts. In another study by Dahal (2005) in the Himalayas of Nepal, some similar impacts were reported. Among the positive impacts, Himalayan community in Nepal also reported easier life in winter due to warmer weather. Other positive impacts reported better tasting apples, longer growing season and the possibility to grow new vegetables without plastic-houses due to higher temperature. Also, more profits from tourism due to longer drought in post-monsoon season were reported. However many people reported negative impacts like destruction of mud roofs and walls due to erratic monsoon, and higher incidence of pests and diseases in crops. It was found that the impacts felt by the people depended on their sources of livelihoods; while those depending on apple orchards and tourism were enjoying positive impacts, those depending on staple crops were suffering due to more pests and diseases. On the other hand, a study conducted in the lowlands of Nepal (Maharjan et al. 2011) does not report any positive perceptions regarding the impacts of extreme events brought about by changes in climate. The major impacts reported were increased pests and diseases in crops and livestock, change in cropping pattern, declining productivity, decreasing grazing lands, and destruction of infrastructure due to higher flood incidences.

10.4 A Case Study of Chepang Community in Rural Mid-Hills of Nepal

The government of Nepal has identified 59 ethnic groups as the indigenous nationalities of Nepal. According to National Foundation for Development of Indigenous Nationalities Act 2002, the term indigenous nationalities in Nepal refer to tribes or communities having their own mother language and traditional rites and customs, distinct cultural identity, distinct social structure and written or unwritten history. Based on the same Act, Nepal Government has identified 59 Indigenous Nationalities who are classified into five groups comprising of endangered, highly marginalized, marginalized, disadvantaged, and advanced group based on a composite index comprising of variables like literacy rate, housing, land holdings, occupation, language, graduates, residence, and population size. The indigenous nationalities are further classified into Mountains, Hills and Tarai based on the geographical location where they form a majority (NIRS 2006). Chepangs are one of the indigenous nationalities of Nepal having a population of 52,237 constituting 0.23 % of the total population. The majority of the Chepangs lives in the hilly villages of Chitwan, Makwanpur, Dhading and Gorkha districts. In Nepal, indigenous nationalities represent the marginalized section of the country. Their socio-economic and human development indicators lie far below the national average. Based on the Nepal Living Standards Survey 2003/04, hilly indigenous people (besides Newar and Thakali) have a higher poverty incidence of 43 % compared to the Tarai indigenous people having a poverty incidence of 33 % (NIRS 2006). Newar and Thakali have been separated from the rest of the hilly indigenous nationalities because they are the only two indigenous nationalities falling under the advanced category, thus have fared much better in the socio-economic indicators compared to others. The Chepang community has been categorized as one of the highly marginalized indigenous nationalities from the hills. Although their native area is surrounded by major highways of the country and is situated very near to the capital city Kathmandu (Fig. 10.1), they are still marginalized from the mainstream of development of the country. Chepangs, thus, qualify as an appropriate representative of the marginalized group of people in Nepal and is, thus, selected as the population for this study.

Chepangs are believed to be, until the last 100–150 years ago, a nomadic group ranging the forests of Nepal as described by Brian Hodgson in his 1848 article to be "living entirely upon wild fruit and the produce of the chase" (Hodgson 1874, p. 45). Nearly a century later, a comprehensive study by Rai (1985) reported that, though Chepangs still practiced a good deal of hunting and gathering, agriculture formed the mainstay of their livelihood, and they practiced khoriya (shifting culture) cultivation. Under this system, a patch of land is cleared in the forest and cultivated for 2–3 years before the soil become exhausted. It is then left fallow allowing sufficient time for vegetation to regenerate; meanwhile they clear and cultivate other patches of land. However, the introduction of new government policies puts restrictions on hunting, gathering, and clearing of forest patches (Upreti and Adhikari 2006), leading to the transition of their livelihood to sedentary agriculture (FORWARD 2001). Chepangs predominantly rely on rain-fed subsistence agriculture. Due to rugged topography and stony nature of the land, only a small percentage of Chepang households are

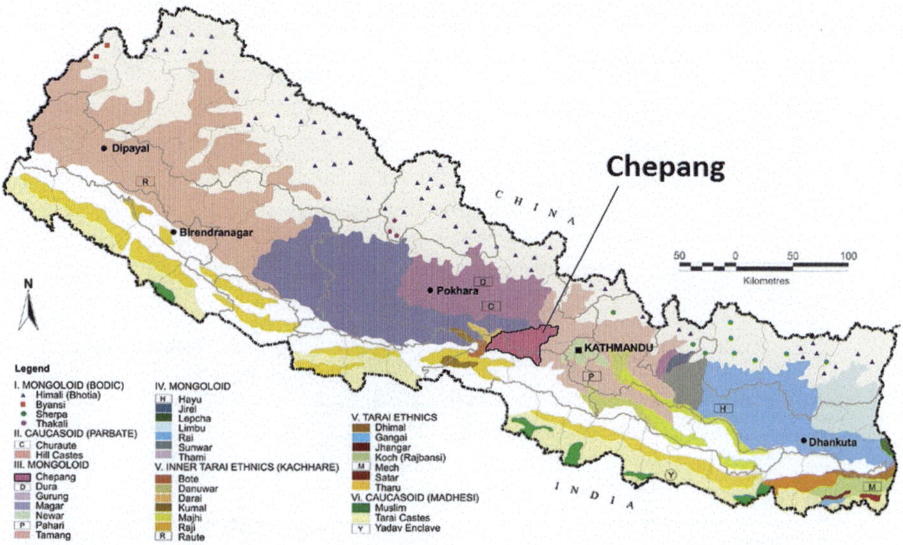

Fig. 10.1 Ethnographic map of Nepal highlighting native area of Chepangs (Turin 2007)

fully food self-sufficient (Piya et al. 2011a). Chepangs still depend upon wild and uncultivated edible plants (Maharjan et al. 2010). Chepangs also depend upon livestock, wage laboring, collection and sale of forest products, handicrafts, skilled non-farm jobs, salaried jobs, and remittance for cash income (Piya et al. 2011b). During the time of their food deficit, the Chepangs depend on informal sources for loans, which they pay back by selling their livestock, and agricultural or forest products.

10.4.1 Study Area and Data Source

This study covers all four districts that form the native area of the Chepangs, i.e. Chitwan, Makwanpur, Dhading and Gorkha districts. One Village Development Committee (VDC) from each district is selected based on the dominance of Chepang population. Kaule VDC from Chitwan district, Kankada VDC from Makwanpur district, Mahadevsthan VDC from Dhading district, and Bhumlichowk VDC from Gorkha district form the four study VDCs (Fig. 10.2). VDCs are the lowest administrative tiers in Nepal, composed of nine wards. The Chepang settlements on hill tops are scattered and connected by narrow foot-trails. One Chepang settlement is separated from the other by a rivulet that flow in the grove between the ridges so that in order to go from one settlement to another, one has to climb down the grove, cross the rivulet, and again climb up the ridge. During monsoon, the rivulets are flooded, and the ridges are very slippery so that movements across the settlements become very difficult; the trails are covered by bushes with plenty of leeches, and falling stones with constant danger of landslides.

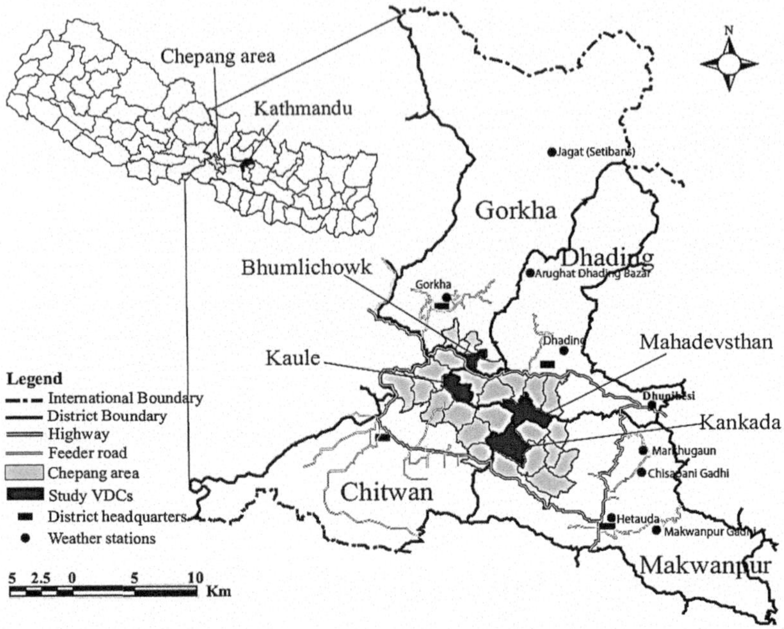

Fig. 10.2 Map of study districts showing sample VDCs (Piya et al. 2012)

This study is based on the primary data collected by household survey conducted in two phases. The field visits, research sites, and sample size have been summarized in Table 10.1.

The first phase of household survey was conducted in February–March 2010 and the second phase in May–June 2011. Sixty randomly selected households from each VDC from the sample for the household survey. Household survey was conducted using semi-structured interview schedule. The researcher visited the selected households and face-to-face interviews were conducted with the household members on the selected household's premises. All the households covered by the survey were untouched by roads and not connected to the central electricity grid. The first phase of the household survey was focused on collection of data related to demographics, livelihood assets, livelihood activities, income sources, and expenditures. Besides the household survey, group discussions were carried out to assess the general changes in climate variables and obtain a timeline of climate related disasters as far as they could remember. Based on the overall general information obtained from the group discussion in 2010, semi-structured interview schedule was designed and follow-up field visit was again made in May–June 2011. This time the same households covered in 2010 were revisited for gathering supplementary data. Out of the total 240 households covered in the 2010 field survey, 58 households in Chitwan, 56 households in Makwanpur, 54 households in Dhading, and 53 households in Gorkha could be revisited in the 2011 survey; thus the final sample constitutes a total of 221 households. The main purpose of this household survey was to find out whether individual households perceived any changes in climate, and if they did, what were the changes

Table 10.1 Field visits, VDCs, and sample size covered by the study (NGIIP 2006a, b; CBS 2002a, b)

District and VDC	Altitudinal range of the VDC (m)	Chepang population in the VDC	Sample size (number of Chepang households) surveyed	
			February–March 2010	May–June 2011 (final sample)
Chitwan—Kaule	810–1,920	3,155 (67.3 %)	60 (11 %)	58 (11 %)
Makwanpur—Kankada	385–1,710	4,056 (52.3 %)	60 (9 %)	56 (8 %)
Dhading—Mahadevsthan	550–1,930	1,857 (30 %)	60 (19 %)	54 (17 %)
Gorkha—Bhumlichowk	410–1,730	911 (24.3 %)	60 (39 %)	53 (35 %)

Note: Figures in parentheses indicate percentage of the VDC total

perceived, adaptation strategies adopted, and the impacts felt on crop production and livelihood assets. A time frame of the past 10 years was considered since a longer time frame would be difficult for the respondents to remember and be subjected to recall bias (Gbetibouo 2009). Also, the year 2001 was taken as the reference year because in July 2001, a large landslide in Kankada VDC claimed more than 60 human lives and caused huge property damage. The Chepangs in all the districts are aware of this epoch making incident, thus it becomes easier for taking this incident as a reference.

For comparing the community perceptions with the actual climate data, historical weather data comprising of mean monthly maximum temperature, minimum temperature, and rainfall were obtained for the year 1975–2008 from the Department of Hydrology and Meteorology (DHM) in Kathmandu, Nepal. Unfortunately, there are no weather stations located in any of the study VDCs. Thus, other weather stations situated at a comparable altitudinal range within the study districts were selected for the study purpose. Four weather stations in the Makwanpur district (Chisapanigadhi, Hetauda, Markhugaun, Makwanpurgadhi), three in Dhading district (Arughat, Dhading, Dhunibesi), and two in Gorkha district (Jagat-setibans, Gorkha) were selected (Table 10.2). Rainfall data are available for all these stations, whereas temperature data is available only from one station in each district (Hetauda, Dhunibesi, and Gorkha). As for Chitwan district, all the weather stations were located below 300 meters above sea level (masl), which is far below the altitudinal range of the study VDC (810–1,920 masl). Therefore, none of the weather stations were considered from Chitwan district. This poses limitations for triangulating the perceptions of respondents from Chitwan district due to lack of recorded climate data for the given elevation range within the district.

10.4.2 Awareness About Climate Change in the Chepang Community

When asked if they have heard about climate change only 11.8 % of the respondents replied positively. When further asked if they can say what climate change means, only 4.5 % of the respondents could reply that the phenomenon is related to changes in weather patterns, temperature, rainfall, wind, floods, landslides, and the environment. The source of information was cited as radio by 6.9 %, staffs of

Table 10.2 Weather stations selected for the purpose of the study

District	Weather stations	Altitude (m)[a]	Available data
Chitwan	None available within the similar altitudinal range as the study VDC		
Makwanpur	Chisapaani Gadhi	1,707	Precipitation (1970–2008)
	Hetauda	474	Precipitation (1966–2008) Temperature (1966–2008)
	Markhugaun	1,530	Precipitation (1972–2008)
	Makwanpur Gadhi	1,030	Precipitation (1975–2008)
Dhading	Arughat Dhading Bazar	518	Precipitation (1971–2008)
	Dhading	1,420	Precipitation (1970–2008)
	Dhunibeshi	1,085	Precipitation (1971–2008) Temperature (1975–2008)
Gorkha	Jagat (Setibans)	1,334	Precipitation (1971–2008)
	Gorkha	1,097	Precipitation (1971–2008) Temperature (1971–2008)

[a]Raw data from DHM

non-governmental organizations (NGOs)[1] by 2.5 % and teachers at school by 1.5 % of the respondents. A similar situation is reported by Byg and Salick (2009) in Tibet where the respondents have never heard the term climate change. Most of the Chepangs may not literally understand what climate change means; but many of them can perceive how weather pattern has varied over the years. They have experienced, for instance, that the rainfall patterns are changing, winter and post-winter rains are decreasing, monsoon is arriving late but causing lots of damage when it arrives, summer is getting hotter, hailstorms are increasing in frequency, and so on. This section presents an overall view of how the study community perceives the ongoing changes in climate. Trends of temperature and rainfall are presented as graphs side by side to community perceptions to see if the perceptions really match with the actual trends. Trend analysis has been done for two different time periods, the long-term trend for 1975–2008 and short-term trend for the time period of 2001–2008. The latter period was chosen since our household survey was more focused on the perceptions based on last 10 years (2001–2010). Rural households tend to form their perception based on more recent events (Maddison 2007), thus the community perception is believed to be more representative of the climatic patterns after 2001. As stated in the methodology section, there are no weather stations at the suitable elevation within the Chitwan district; this puts limitations in the triangulation of the perception of respondents from Chitwan district. The comparison of perceptions with that of recorded data is possible only for the remaining three districts.

[1]The major NGOs who have worked or are currently working in the study areas are Support Activities for Poor Producers of Nepal (SAPPROS), Forum for Rural Welfare and Agricultural Reform for Development (FORWARD), Local Initiatives for Biodiversity Research and Development (LI-BIRD), Focus Nepal, Shanti Nepal, Center for Community Development, Nepal (CCDN), Center for Community Development and Research (CCODER), Manahari Development Institute (MDI), and Practical Action Nepal. These organizations provides rural community development services mostly in the sector of agriculture, livestock, natural resource management and conservation, drinking water, community health, savings and credits, small-scale irrigation, and renewable energy (micro-hydro, solar lighting systems, and improved cooking systems).

Table 10.3 Perceptions of changes in temperature (field survey, 2010/2011)

	Number of response				
Perceptions	Aggregate (n=221)	Chitwan (n=58)	Makwanpur (n=56)	Dhading (n=54)	Gorkha (n=53)
Hotter summer	30 (13.6)	5 (8.6)	9 (16.1)	8 (14.8)	8 (15.1)
Cooler summer	4 (1.8)	2 (3.4)	2 (3.6)	–	–
Colder winter	3 (1.4)	–	–	2 (3.7)	1 (1.9)
Warmer winter	3 (1.4)	–	2 (3.6)	–	1 (1.9)
Hotter summer and colder winter	38 (17.2)	–	7 (12.5)	15 (27.8)	14 (26.4)
Hotter summer and warmer winter	37 (16.7)	15 (25.9)	3 (5.4)	6 (11.1)	13 (24.5)
Cooler summer and colder winter	7 (3.2)	2 (3.4)	4 (7.1)	1 (1.9)	–
Cooler summer and warmer winter	10 (4.5)	3 (5.2)	4 (7.1)	1 (1.9)	2 (3.8)
Fluctuating between the years	1 (0.5)	1 (1.7)	–	–	–
No changes perceived	85 (38.5)	26 (44.8)	25 (44.6)	20 (37.0)	14 (26.4)
Don't know	3 (1.4)	2 (3.4)	–	1 (1.9)	–

Note: Figures in parentheses indicate percentage; n number of sample households

10.4.3 Changes in Temperature: Perceptions and Actual Trends

Regarding the changes in temperature, the majority of respondents has noticed the rising summer temperature (47.5 %), while nearly 10 % of the respondents perceive that summer has become cooler. For the winter temperature, nearly 21.8 % perceive that winter is becoming colder while nearly equal percentage of the respondents (22.6 %) perceive that winter is getting warmer. The perceptions are similar to those reported by other studies done in the hills of Nepal where the majority of respondents perceive an increase in overall temperatures (Bhusal 2009; Tiwari et al. 2010); however these studies do not differentiate seasonal temperatures. Study by Dahal (2005) in the high Himalayas reports that the community perceived winters to be warmer and less frosty. In our study, there are also a significant proportion of the respondents (38.5 %) who do not perceive any changes in temperature. The reason why many respondents cannot perceive long-term changes in temperature might be because of what Vedwan and Rhoades (2001) describe as the lack of "visual salience." According to the authors, visual salience of rainfall facilitates better perception, whereas changes in temperature are comparatively perceived lesser. The detail of the categories of response to temperature changes has been tabulated in Table 10.3.

A district wise response shows that in Chitwan and Makwanpur, there is a quite big proportion (nearly 45 %) of respondents who do not perceive any changes in temperatures; the proportion is 37 % and 26 % for Dhading and Gorkha, respectively. Of the remaining who perceives changes, the majority perceives rising summer temperature (35 %, 34 %, 54 %, and 66 % in Chitwan, Makwanpur, Dhading, and Gorkha, respectively). Regarding the winter temperature, the response is not as uniform and clear as the summer temperature. In Chitwan, the majority (31 %) perceives winter is getting warmer; in Makwanpur nearly 20 % perceive colder winter while 16 % perceive warmer winter; in Dhading the majority (33 %) perceive colder winter; and in Gorkha the two figures are again closer to 30 % perceiving warmer winter and 28 % perceiving colder winter.

a

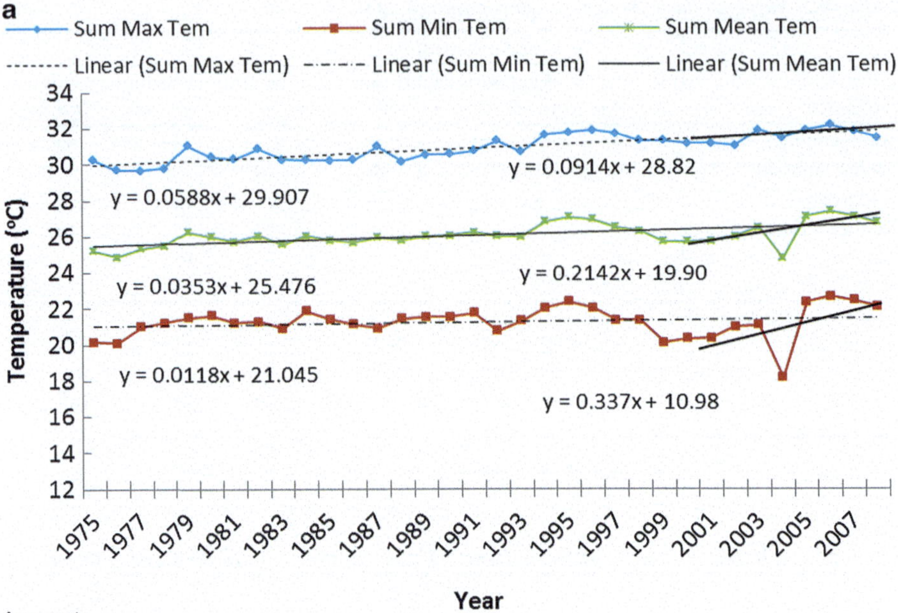

Aggregate

b

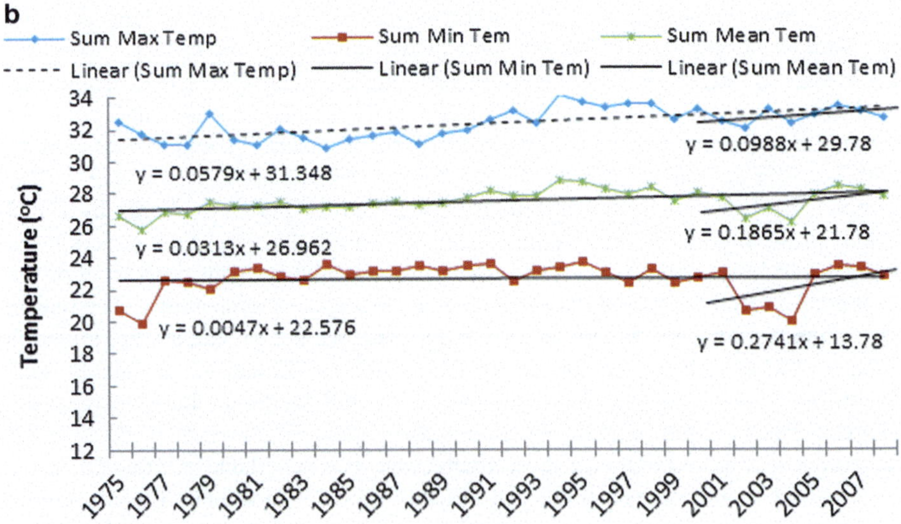

Makwanpur

Fig. 10.3 (**a**) Average summer temperature trend (May–August) for the selected stations in the research districts (raw data from DHM). (**b**) Average summer temperature trend (May–August) for the selected stations in Makwanpur district (raw data from DHM). (**c**) Average summer temperature trend (May–August) for the selected stations in Dhading district (raw data from DHM). (**d**) Average summer temperature trend (May–August) for the selected stations in Gorkha district (raw data from DHM)

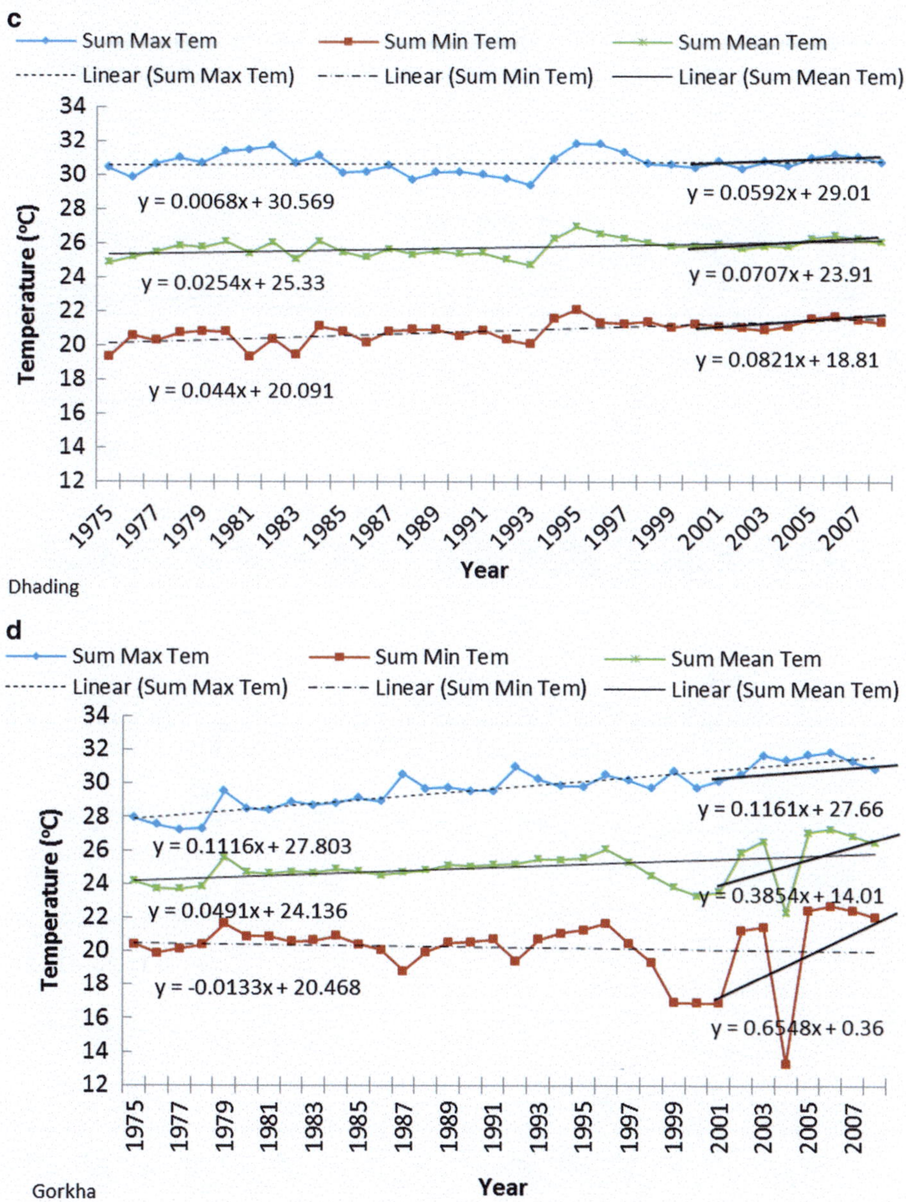

Fig. 10.3 (continued)

The recorded summer temperature trend is shown in Fig. 10.3a–d, and winter temperature trend in Fig. 10.4a–d. The long-term summer temperature (May–August) shows rising trend for all the districts except for the minimum summer

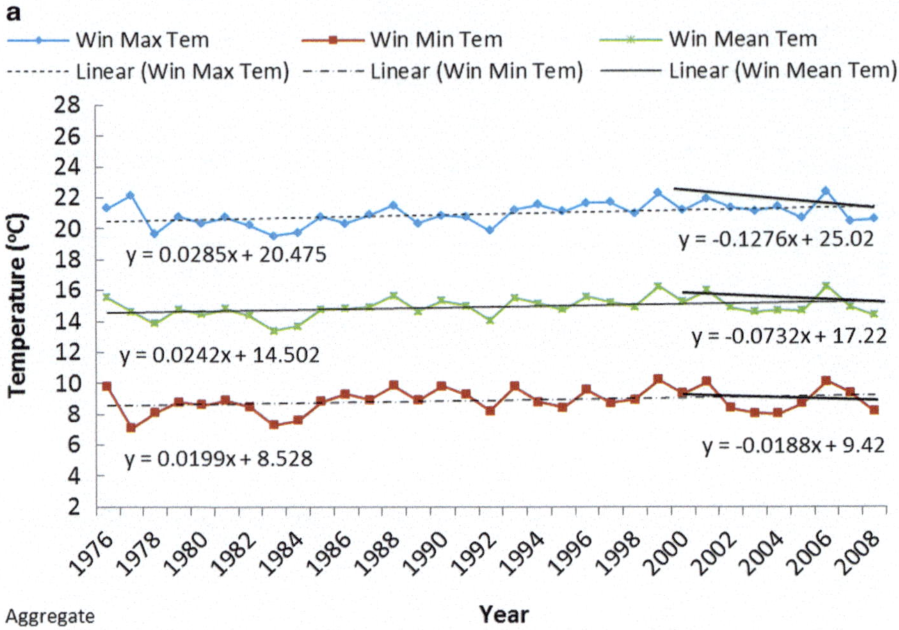

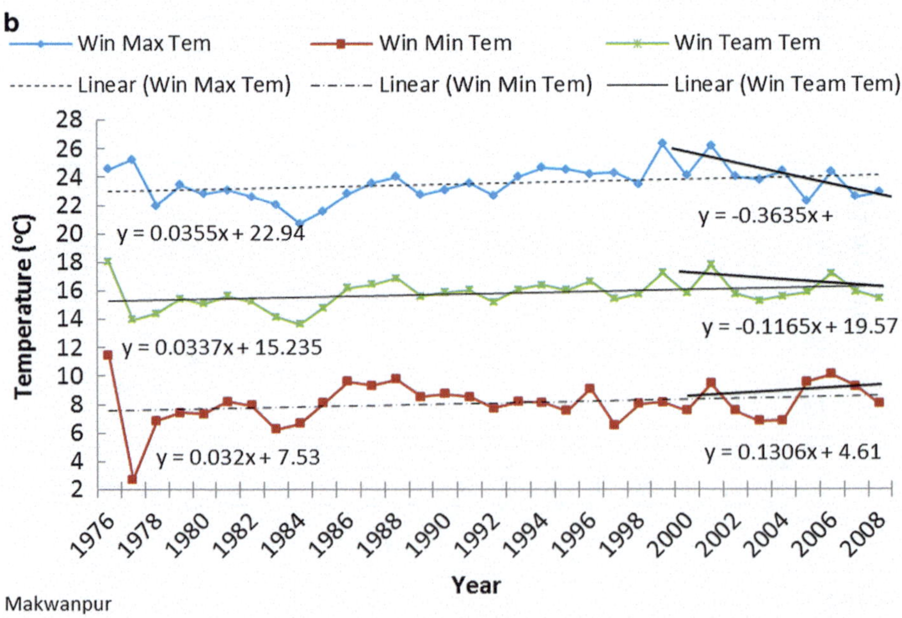

Fig. 10.4 (**a**) Average winter temperature trend (December–February) for the selected stations in the research districts (raw data from DHM). (**b**) Average winter temperature trend (December–February) for the selected stations in Makwanpur district (raw data from DHM). (**c**) Average winter temperature trend (December–February) for the selected stations in Dhading district (raw data from DHM). (**d**) Average winter temperature trend (December–February) for the selected stations in Gorkha district (raw data from DHM)

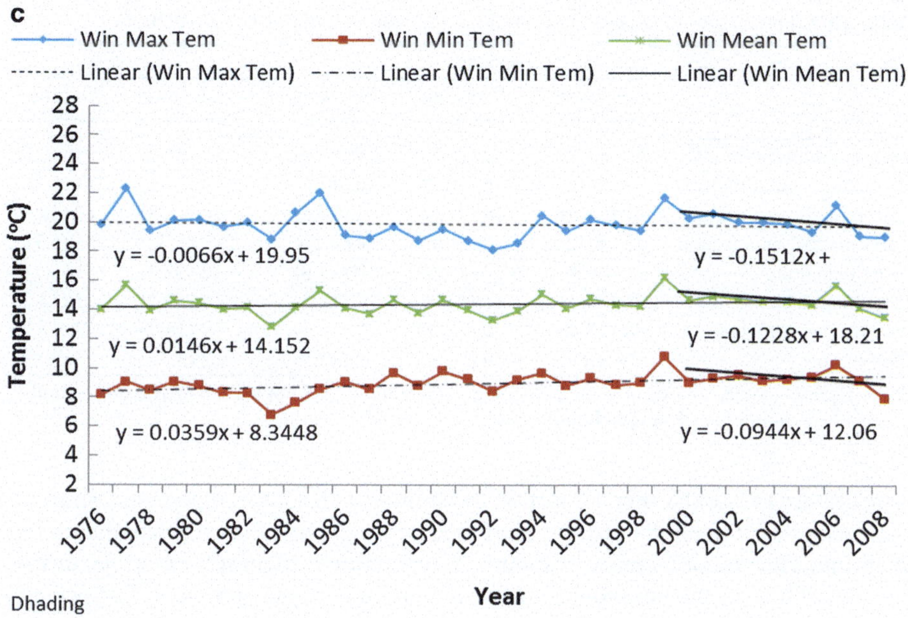

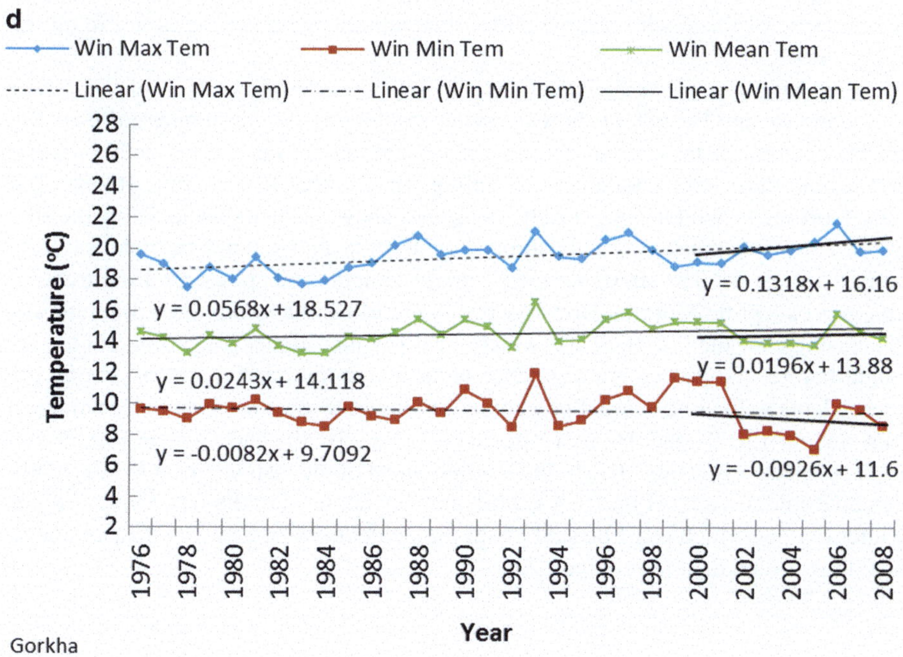

Fig. 10.4 (continued)

Table 10.4 Perceptions of changes in overall rainfall pattern (field survey 2010/2011)

	Number of response				
Perceptions	Aggregate (n=221)	Chitwan (n=58)	Makwanpur (n=56)	Dhading (n=54)	Gorkha (n=53)
Decreased total rainfall	81 (36.7)	12 (20.7)	19 (33.9)	17 (31.5)	33(62.3)
Increased total rainfall	22 (9.9)	6 (10.3)	10 (17.8)	3 (5.6)	3 (5.7)
Late rainfall than usual	47 (21.3)	16 (27.6)	11 (19.6)	10 (18.5)	10 (18.9)
Early rainfall than usual	13 (5.9)	3 (5.2)	4 (7.1)	2 (3.7)	3 (5.7)
More damaging	23 (10.4)	8 (13.8)	10 (17.8)	12 (22.2)	3 (5.7)
Unpredictable	52 (23.5)	25 (43.1)	8 (14.3)	19 (35.2)	7 (13.2)
No changes perceived	34 (15.4)	9 (15.5)	9 (16.1)	9 (16.7)	7 (13.2)
Don't know	9 (4.1)	3 (5.2)	2 (3.6)	2 (3.7)	2 (3.8)

Note: Figures in parentheses indicate percentage; Percentage may not add up to 100 due to multiple answers; n number of sample households

temperature in Gorkha (Fig. 10.4d). In the short-run also, the temperatures show an increasing trend in all the districts, and the rate of increase is faster than that in the long-run. The rate of increase in summer temperature is highest for Gorkha; unsurprisingly 66 % of the households in Gorkha perceive hotter summers. The perceptions regarding summer temperature are rightly perceived in other districts also, as the majority of those who responded felt that summers are getting hotter in all the study sites.

In the long-run, trend analysis of winter temperatures (December of the earlier year, January and February) shows varying results with winter temperatures rising for Makwanpur district, a maximum winter temperature falling for Dhading district and a minimum winter temperature falling for Gorkha district (Fig. 10.4d). The long-term mean winter temperature, however, shows a rising trend in all the districts. However, the trend analysis of short-term winter temperature shows quite surprising results. The short-run mean winter temperature shows a falling trend in all cases except for Gorkha. However even in Gorkha, the minimum temperature is decreasing over the last 10 years, which means winter nights are getting colder. As stated before, the perception regarding winter temperature is not clear among the respondents with almost equal percentage saying both warmer and colder winter in Makwanpur (20 % and 16 %, respectively) (Fig. 10.4b) and Gorkha (30 % and 28 %, respectively) (Fig. 10.4d). In Dhading, most of the respondents (33 %) rightly perceive colder winter (Fig. 10.4c), while in Chitwan most of them perceive warmer winter (31 %). Much can't be said about the perceptions in Chitwan district due to lack of recorded data for comparison.

10.4.4 Changes in Rainfall: Perceptions and Actual Trends

Perception of rainfall was asked to the respondents in terms of rainfall quantity as well as timings (Table 10.4). In terms of quantity, the majority of the respondents perceives a decrease in rainfall and in the terms of timing a majority of the respondents perceive rainfall is arriving later in the year at all of the study sites. This perception is quite similar to what is mentioned in the studies conducted in the Himalayas (Chaudhary and Bawa 2011; Byg and Salick 2009; Vedwan 2006), whereby the community perceived that snowfall and rainfall has shifted to a later in the year. Most of the respondents (37 %) feel that the total rainfall has decreased; the next majority of respondents (24 %) feels that rainfall is unpredictable in terms of quantity (sometimes high, sometimes low); there are also quite many respondents (21 %) who feel that rainfall is coming later than the usual time. Quite many respondents in Chitwan and Dhading perceive the unpredictable nature of rainfall. Other studies in Nepal also report that most respondents perceive rainfall to be very unpredictable regardless of whether the study was conducted in low-lying Tarai, mid-hill, high-hills, or mountains (Tiwari et al. 2010; Bhusal 2009; Dahal 2005). In our study, there are around 13–17 % of respondents who do not perceive any changes in rainfall; yet this is far less compared to those who did not perceive any changes in temperature.

Triangulating the rainfall perceptions with actual trends was difficult, since monthly averages can give the picture of total amount but do not give a picture of rainfall timings. Trends in total annual rainfall are presented in Fig. 10.5a–d. Rainfall patterns show that the interannual variations are very large for all the districts, thereby making the rainfall pattern quite unpredictable. The rainfall pattern for overall Nepal also follows the same pattern of large variations, making it difficult to draw a single conclusion regarding the rainfall patterns (Practical Action 2009). Trend analysis for rainfall was also done for two time periods, the first being the period of

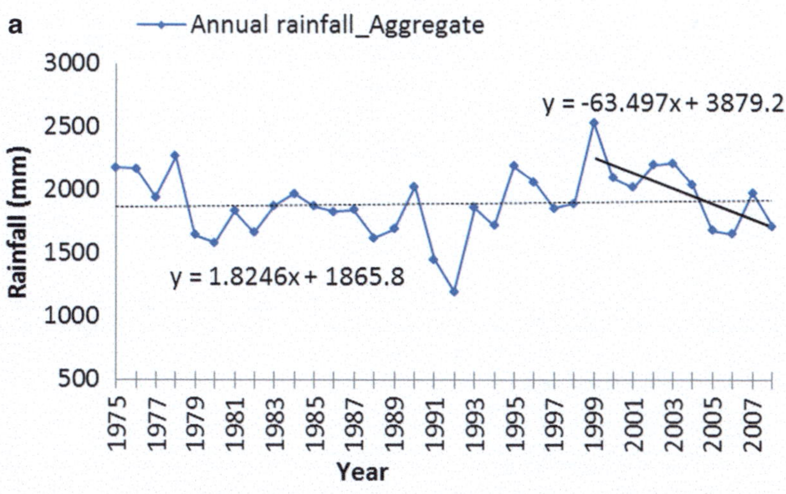

Fig. 10.5 (**a**) Total annual rainfall averaged for the selected stations in the research districts (raw data from DHM). (**b**) Total annual rainfall averaged for the selected stations in Makwanpur district (raw data from DHM). (**c**) Total annual rainfall averaged for the selected stations in the Dhading district (raw data from DHM). (**d**) Total annual rainfall averaged for the selected stations in Gorkha district (raw data from DHM)

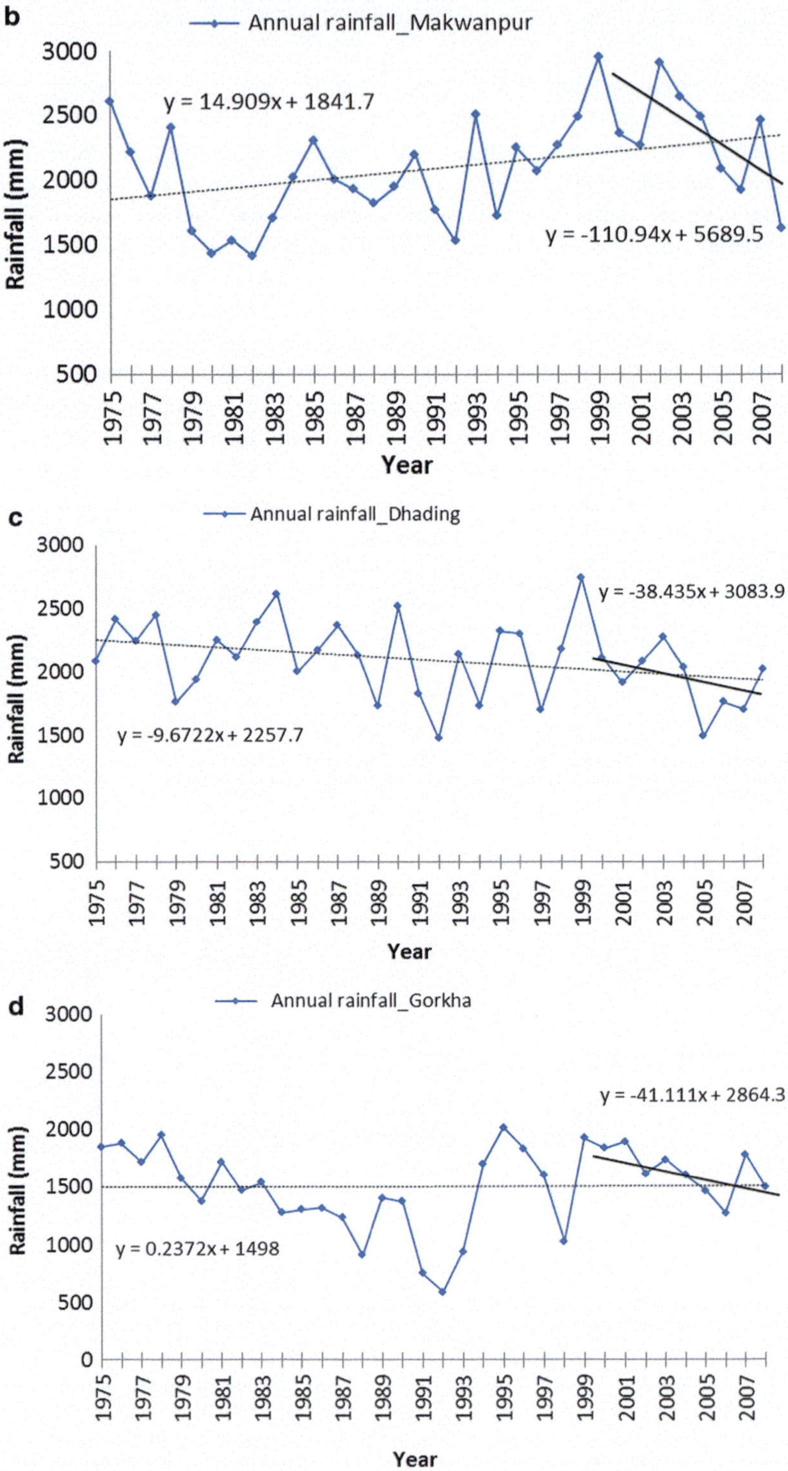

Fig. 10.5 (continued)

Table 10.5 Impacts of changes in temperature (field survey, 2010/2011)

Impacts of	Types of impacts
Higher temperature on crops	Drying of crops (43), more flower drop in fruits (1), yellowing of maize (1)
Higher temperature on livestock	Death of small livestock due to heatstroke (3), more diseases in livestock (13), lesser livestock productivity (6), less fodder availability (7)
Higher temperature on human	Illness in human (diarrhea, vomiting, dysentery, indigestion, loss of appetite, headache) (25), difficult to work (labor) (14), decay of food (1)
Lower temperature on crops	Dews decay millet (1)
Lower temperature on livestock	Death of small livestock due to cold (3)
Lower temperature on human	Illness in human (cough, cold) (3)

Note: Figures in parentheses indicate the number of responses received

1975–2008, and another for the period 2001–2008. Similar to winter temperatures, trend diagrams for annual rainfall also show that the trends for the two different time periods can be totally different. For Makwanpur district, the trend in the long run is seen to be slightly increasing, however, the rainfall amount shows a drastically decreasing trend for the period after 2001 (Fig. 10.5b). For the other two districts also, rainfall after 2001 is decreasing at a faster rate than the overall trend (Fig. 10.5c–d). The actual rainfall trend matches with the community perception, where the majority says that the quantity of rainfall is decreasing and is unpredictable. Among the four study sites, the vast majority of respondents from Gorkha (62 %) could perceive decreasing rainfall in line with what the records show, while the proportion was least in Chitwan (21 %). Alternatively, those perceiving unpredictable nature of rainfall (that is high is some years and low in others) is highest in Chitwan (43 %) and lowest in Gorkha (13 %). Many respondents feel that rainfall pattern is unpredictable probably due to large interannual fluctuations in the rainfall quantity.

10.4.5 Impacts Reported by the Community

After asking the respondents if they perceived any changes in temperature and rainfall, those perceiving some changes were asked about the impacts those changes brought in their livelihoods. The impacts of changes in temperature are summarized in Table 10.5. Corresponding to the majority of respondents perceiving rising temperature, most of the impacts felt are also due to higher temperature. The most cited impact are drying of crops followed by human illnesses like diarrhea, vomiting, indigestion, dysentery, loss of appetite and headache due to higher temperature, while one of the respondents also said that the foods decay due to higher temperatures. Many respondents also reported difficulties to work during summer, and higher incidences of diseases like bloating in the livestock. Responses of lesser fodder availability, lesser livestock productivity, and death of small livestock (especially poultry) due to heat stroke are also obtained. The impacts felt due to lower

Table 10.6 Impacts of changes in rainfall (field survey, 2010/2011)

Impacts of	Types of impacts
Lesser rainfall on crops	Less production due to less/no rain (53), drying of crops/fruit trees (26)
Lesser rainfall on livestock	Lesser fodder availability (9), diseases in livestock (9), lesser livestock productivity (3)
Lesser rainfall on human	More illness in human (4)
Shifts in rainfall patterns on crops	Cannot sow/transplant on time (22), less crop production due to untimely sowing (1), hampers land preparation for sowing (3)
Unpredictable rainfall on crops	Alternate dry and wet periods hamper crops (6)
Higher rainfall on crops	Lodging (2), lesser crop production due to water logging (2), more diseases in fruits/crops (3), higher crop production (11)
Higher rainfall on livestock	More fodder available (2)
High rainfall on properties	Properties/crops washed away by heavy rains/landslides (8)

Note: Figures in parentheses indicate the number of responses received

temperature is quite few because the proportion of respondents perceiving decreasing temperature is also lesser. The impacts of lower temperature reported are decaying of millet due to higher dew formation at the time of harvest, death of small livestock due to coldness, and human illnesses due to cold.

The impacts of changes in rainfall quantity and patterns are summarized in Table 10.6. As the majority of respondents perceived decreasing rainfall quantity, many respondents felt that the crop production is declining due to lesser rain or no rain especially during the tasseling and silking stage of maize. Many also reported drying of crops and fruit trees due to lack of rain. Similarly, the lesser fodder availability due to drought was also reported to subsequently result in lesser livestock productivity and diseases in livestock. Human illnesses like fever and headache was also reported. The impact of late rainfall shows the impact on the shift on sowing and transplanting time of crops, one respondent felt crop production is decreasing due to untimely sowing, while a few said that land preparation for sowing maize is hampered due to late pre-monsoon rains. The unpredictable nature of rains is also affecting the productivity of crops. More recently, drought during June to early July is hampering the maize while too much rain in August is hampering the tomato harvest. There are mixed responses obtained from those perceiving higher rainfall. Some respondents said the crop production is higher and fodder is more available due to higher rainfall, while few of them said that too much rain causes lodging of crops, problems of water logging, and higher incidences of crop diseases. Similarly, eight respondents report washing away of standing crops, and land by erratic rainfall during monsoon.

10.5 Conclusion

National level polls reveal that the level of awareness about the phenomenon of climate change is quite low in developing countries. However, among those who are aware, a higher proportion of respondents in the developing countries are concerned about the associated risks and impacts. The residents of developed countries are less worried about the possible adverse impacts of climate change. This demonstrates that unless the public sees the impacts happening in their own locality, it is difficult to raise the level of concern among them. Among the developed countries, the US was seen as the least concerned country and also least supportive of mitigation measures. The US government also seems to be in line with the public opinion as it has not included itself within the Kyoto Protocol, despite being one of the largest emitters in the world. Although the aware public in all the countries agree widely that there is a need to undertake necessary mitigation actions, a majority of the public in both developing as well as developed countries are not in favor of mitigation policies and actions that are directly related to increased personal spending. This hints towards a serious challenge in formulating and implementing climate change mitigation policies around the globe.

Public awareness raising is important in both developed as well as developing countries. In developed countries, the people need to know that climate change has very serious ramifications for the residents in developing countries, and they need to be convinced that small individual contributions from their side to reduce the energy use can be a significant contribution in climate change mitigation. In the developing countries, awareness is necessary in order to design proactive adaptation measures at the local level so that the climate change impacts can be minimized. The governments in all countries should put climate change issues in the priority list of policies, and mainstream climate change in integration with all other development sectors like infrastructure development, agriculture, health, and so on.

At the local level, the trend analysis of temperature and precipitation trend for both long-term and short-term provides some important insights. Firstly, the direction of trends can differ for the two time-periods, as shown by the trends for winter temperature and annual rainfall. Secondly, as seen in the case of rainfall, community perceptions are more in line with short-term trends, rather than with the long-term trends. It is the latest trend that has effects on the people's livelihoods directly and the decisions taken to adapt accordingly. Policy makers should be critical to analyze both the long term as well as the short term trends, before implementing any development decisions.

Around one-third of the respondents in our case study perceive the changes in line with the data recorded in the weather stations. A matter of concern is that there is a significant proportion of the population who has not been able to perceive any of those changes. The scenario is similar to most of the rural farming communities thereby calling for a need for awareness raising and information dissemination in these rural areas, where the livelihoods are predominantly dependent upon farming. Unless the community realizes that there have been changes going on in the weather patterns, they cannot be motivated to take appropriate measures to adapt to their farming systems according to these changes. Studies have shown that information dissemination and community level extension services are very effective to inform the people about such changes and to convince them to take necessary adaptation actions. The NGOs, and local governments working at the grassroots level can play a very important role in disseminating the relevant information and conducting awareness raising campaigns. Rural communities need information about seasonal weather forecasts and they should be assisted to design their crop calendar in accordance with these forecasts. Only broadcasting such information through radio and television is not sufficient as not all the rural households possess radio or television; and their time schedule may not be flexible enough to listen to those broadcasts. These types of information will be more effective if broadcasted by extension agents through direct interaction with the community and such programs should emphasize participation of both male and female.

Finally, monitoring climate changes in rural areas in developing countries is hindered by the lack of sufficient weather stations and recorded data. An establishment of small hydrological stations at the local level in rural areas is recommendable. As already recommended by Dahal (2005), training the staffs and students at local schools or members of local community-based organizations to obtain readings from rain-gauge and thermometers would not only make it possible to generate datasets on local climate, but it would also be easier to raise awareness among the local communities about the changing climate, and the appropriate measures that can be taken to tackle its adverse impacts.

References

BBC (2010) BBC climate change poll—February 2010. Retrieved 29 December 2011, from news.bbc.co.uk/nol/shared/bsp/hi/pdfs/05_02_10climatechange.pdf

Bhusal YR (2009) Local people's perceptions on climate change, its impacts and adaptation measures in Mid-Mountain region of Nepal (A case study from Kaski district). Unpublished BSc forestry thesis. Institute of Forestry, Pokhara

Byg A, Salick J (2009) Local perspectives on a global phenomenon—climate change in eastern Tibetan villages. Global Environ Change 19:156–166

CBS (2002a) Population of Nepal. Village development committees/municipalities. Population census 2001—selected tables on caste/ethnicity, mother tongue, and religion (central development region). National Planning Commission Secreteriat, Central Bureau of Statistics (CBS), Kathmandu

CBS (2002b) Population of Nepal. Village development committees/municipalities. Population census 2001—selected tables on caste/ethnicity, mother tongue, and religion (western development region). National Planning Commission Secreteriat, Central Bureau of Statistics (CBS), Kathmandu

Chaudhary P, Bawa KS (2011) Local perceptions of climate change validated by scientific evidence in the Himalayas. Retrieved 28 Nov 2011, from Biology Letters: http://harvardforest.fas.harvard.edu/publications/pdfs/Chaudhary_BioLetters_2011.pdf

Dahal N (2005) Perceptions in the Himalayas. Tiempo 56:19–24

Deressa TT, Hassan RM, Ringler C (2011) Perception of and adaptation to climate change by farmers in the Nile basin of Ethiopia. J Agric Sci 149:23–31

FORWARD (2001) Formulation of a comprehensive praja development programme, part II, Program development. Prepared for Ministry of local development, Kathmandu, and SNV-Nepal. Forum for Rural Welfare and Agricultural Reform for Development (FORWARD), Chitwan

Gbetibouo GA (2009) Understanding farmers' perceptions and adaptations to climate change and variability. The case of the Limpopo basin, South Africa. IFPRI discussion paper 00849. International Food Policy Research Institute, Washington, DC

GlobeScan (2006) 30-Country poll finds worldwide consensus that climate change is a serious problem. Retrieved 29 December 2011, from www.globescan.com/news-archives/csr_climatechange.html

Hodgson BH (1874) On the Chepang and Kusunda tribes of Nepa. In: Essays on the languages, literature and religion of Nepal and Tibet. Manjushri Publishing House, New Delhi, pp 45–54

IPCC (2007) Climate change 2007: impacts, adaptation, and vulnerability. Contribution of working group II to the fourth assessment report of the intergovernmental panel on climate change. Cambridge University Press, Cambridge

Kim SY (2011) Public perceptions of climate change and support for climate policies in Asia: evidence from recent polls. J Asian Stud 70(2):1–13

Leiserowitz A (2006) Climate change risk perception and policy preferences: the role of affect, imagery, and values. Climatic Change 77:45–72

Leiserowitz A (2007) Public perception, opinion and understanding of climate change—current patterns, trends and limitations. United Nations Development Programme, New York

Lorenzoni I, Pidgeon NF (2006) Public views on climate change: European and USA perspectives. Climatic Change 77:73–95

Maddison D (2007) The perception of and adaptation to climate change in Africa. Policy research working paper 4308. The World Bank, Washington, DC

Maharjan KL, Piya L, Joshi NP (2010) Annual subsistence cycle of Chepangs in the Mid-Hills of Nepal: an integration of farming and gathering. Himalayan J Sociol Anthropol 4:105–133

Maharjan SK, Sigdel ER, Sthapit BR, Regmi BR (2011) Tharu community's perception on climate changes and their adaptive initiations to withstand its impact in Western Terai of Nepal. Int NGO J 6(2):35–42

NGIIP (2006a) Index of geographical names of Nepal, Central development region, vol. II. Retrieved 9 January 2011, from National Geographic Information Infrastructure Programme (NGIIP), Survey Department, Government of Nepal: http://www.ngiip.gov.np/publications/volume2link.php

NGIIP (2006b) Index of geographical names of Nepal, Western development region, vol. III. Retrieved 9 January, 2011, from National Geographic Information Infrastructure Programme (NGIIP), Survey Department, Government of Nepal: http://www.ngiip.gov.np/publications/volume3link.php

NIRS (2006) Socio-economic status of indigenous peoples (based on Nepal Living Standards Survey—2003/04 data set). Nepal Integrated Research System Pvt. Ltd. (NIRS), Kathmandu

Patino L, Gauthier DA (2009) Integrating local perspective into climate change decision making in rural areas of the Canadian prairies. Int J Climate Change Strateg Manage 1(2):179–196

Pew (2006) No global warming alarm in the U.S., China. Retrieved 11 January 2012, from http://www.pewglobal.org/files/pdf/252.pdf

Pew (2011) Modest rise in number saying there is "solid evidence" of global warming. The Pew Research Center, Washington, DC

Piya L, Maharjan KL, Joshi NP (2011a) Forest and food security of indigenous people: a case of Chepangs in Nepal. J Int Dev Cooperation 17(1):113–135

Piya L, Maharjan KL, Joshi NP (2011b) Livelihood strategies of indigenous nationalities in Nepal: a case of Chepangs. J Int Dev Cooperation 17(2):99–114

Piya L, Maharjan KL, Joshi NP (2012) Perceptions and realities of climate change among the Chepang communities in rural mid-hills of Nepal. J Contemp India Stud: Space Soc 2:35–50

Practical Action (2009) Temporal and spatial variability of climate change over Nepal (1976–2005). Practical Action, Kathmandu

Rai NK (1985) People of the stones: the Chepangs of central Nepal. Centre for Nepal and Asian Studies, Tribhuvan University, Kirtipur

Smit B, Wandel J (2006) Adaptation, adaptive capacity and vulnerability. Global Environ Change 16:282–292

Tiwari KR, Awasthi KD, Balla MK, Sitaula BK (2010). Local people's perception on climate change, its impact and adaptation practices in Himalaya to Terai regions of Nepal. Retrieved 29c December 2011, from Nepal Study Center, Himalayan Research papers Archive, The University of New Mexico: http://repository.unm.edu/bitstream/handle/1928/11324

Turin M (2007) Harka's maps. Retrieved 11 July 2011, from Nepali Times: http://www.nepalitimes.com/issue/333/Nation/13160

Upreti BR, Adhikari J (2006) A case study on marginalized indigenous communities' access to natural resources in Nepal: national laws, policies, and practices. Preliminary draft presented at the national thematic dialogue held on 17 February 2006 and 19 February 2006. Kathmandu

Vedwan N (2006) Culture, climate and the environment: local knowledge and perception of climate change among apple growers in Northwestern India. J Ecol Anthropol 10:4–18

Vedwan N, Rhoades RE (2001) Climate change in the western Himalayas of India: a study of local perception and response. Climate Res 19:109–117

WB (2010) Public attitude toward climate change: findings from a multi-country poll. The World Bank (WB), Washington, DC

About the Authors

Keshav Lall Maharjan (Dr. of Agriculture in Agricultural Economics, Kyoto University, Japan) is currently a professor at the Graduate School for International Development and Cooperation, Hiroshima University, Japan, where he has been teaching, conducting research, and chairing various steering and decision-making committees since its foundation in 1995. He gives lectures for graduate students on subjects including rural economics, south Asian studies, international development, and cooperation studies. He conducts weekly seminars at the graduate school that address pertinent issues in agricultural economics, rural development, sustainable development, cultural dynamics, climate change, and rural livelihood strategies in developing countries. He also offers support for graduate students writing their Master's theses and doctoral dissertations on the related topics of development sciences, educational development, and cultural and regional studies, which include issues concerning natural resource management, food security, poverty dynamics, local governance, rural society, and community dynamics. In doing so he considers agriculture and rural regions as not merely sources of cheap labor, cheap food and cheaper intermediate inputs, and subordinate to urban areas and centers as marginal and peripheral regions but also representing a dignified way of life for people who are guardians of nature and are more conscious about the earth, humans, and their interaction, so as to sustain this culture and civilization for generations hereafter. Rural regions are the places that make these things happen. Hence, fieldwork to grasp the diverse realities of rural regions location-specifically before generalizing the research is given importance in his research, lectures, and his work with graduate students in terms of their research, writing journal articles and dissertations. Some 20 students have received their Ph.D. from Hiroshima University under his guidance.

In order to disseminate research findings, consolidate ideas and concepts, and share knowledge with other professionals, he regularly participates in local, national, and international seminars and conferences organized by academic societies, research institutions, and various organizations and like-minded individuals,

K.L. Maharjan and N.P. Joshi, *Climate Change, Agriculture and Rural Livelihoods in Developing Countries*, Advances in Asian Human-Environmental Research, DOI 10.1007/978-4-431-54343-5, © Springer Japan 2013 including agricultural economists, ruralogists, sociologists, environmentalists, anthropologists, educationalists, policy makers, development practitioners, farmers, social activists, local leaders, and opinion shapers.

Some of his earlier books in English include *Peasantry in Nepal: A Study on Subsistence Farmers and Their Activities Pertaining to Food Security*, Hiroshima: Research Center for Regional Geography, Hiroshima University, 2003 and *Impacts of Irrigation and Drainage Schemes on Rural Economic Activities in Bangladesh*, Hiroshima: Research Center for Regional Geography, Hiroshima University, 1997. He has also contributed chapters to publications including *Climate Change: Asian Perspective*, Jaipur: Rawat Publication, 2012; *Public Policy and Local Development—Opportunities and Constraints*, International Geographical Union Commission on Geography and Public Policy, 2008; *Political and Social Transformation in North India and Nepal*, New Delhi: Manohar Publishers, 2007; *Small-Scale Livelihoods and Natural Resource Management in Marginal Areas of Monsoon Asia*, Dehra Dun: Bishen Singh Mahendra Pal Singh, 2006; *New Challenges Facing Asian Agriculture under Globalization*, Selangor: Malaysian Agricultural Economics Association, 2005; *Translating Development: The Case of Nepal*, New Delhi: Social Science Press, 2003; and *Sustainable Agriculture, Poverty and Food Security*, Jaipur: Rawat Publications, 2002. He contributes to various related academic journals and has more than one hundred blind reviewed articles to his credit. He has also produced numerous books and journal articles in Japanese.

Niraj Prakash Joshi, Ph.D., is currently working as a research coordinator (associate professor) at Nepal Engineering College's Centre for Postgraduate Studies (*nec*-CPS) and is responsible for guiding postgraduate students in their research in the fields of natural resource management, integrated water resource management, construction management and transportation engineering and management. He received his Ph.D. in rural/agricultural economics from Hiroshima University in 2011 and served the university as an assistant professor from 2011 to 2012 after having been a research assistant there from 2008 to 2011. During his tenure at Hiroshima he has above all been responsible for advising students on research project development and implementation and on presenting and writing academic reports. He is also intensively engaged in research related to poverty, food insecurity, rural livelihoods and climate change in developing countries, with a particular focus on the poverty-ridden region of the far-western rural hills of Nepal, as well as the impoverished and marginalized community Chepang in Nepal's central remote hills. These research activities directly relied on quantitative as well as qualitative data collected through field surveys, which he was directly involved in right from the initial stage of research project development. Hence, he has gained key experience in the course of various research internships, as well as in academic and independent research projects funded by various international research funding agencies, experience he has used to publish 18 research articles. He has presented papers related to poverty, food insecurity, rural livelihoods and climate change at several national and international conferences and has gathered further direct field experience in the other remote districts of Nepal, conducting mid-term assessments for development projects implemented by government agencies (Directorate of Livestock—poultry project for a self-employment generation and poverty reduction in far and midwestern development region) and nongovernment agencies (Helvetas—benefit cost analysis of Helvetas project in western and midwestern development region). He has also served as a socio-economic advisor in the development of four different projects focusing on bananas, goats, buffalo and citrus fruits in different parts of Nepal in 2005, which were funded by the Nepal Agriculture Research Development Fund.

Index

K.L. Maharjan and N.P. Joshi, *Climate Change, Agriculture and Rural Livelihoods in Developing Countries*, Advances in Asian Human-Environmental Research,
DOI 10.1007/978-4-431-54343-5,

D

The manufacturer's authorised representative in the EU is Springer Nature Customer Service Centre GmbH, Europaplatz 3, 69115 Heidelberg, Germany. If you have any concerns regarding our products, please contact ProductSafety@springernature.com

Printed and bound by CPI Group (UK) Ltd, Croydon, CR0 4YY
25/09/2026
02222841-0014